John Kingsley

Elements of comparative zoology

John Kingsley

Elements of comparative zoology

ISBN/EAN: 9783337277031

Printed in Europe, USA, Canada, Australia, Japan

Cover: Foto ©berggeist007 / pixelio.de

More available books at **www.hansebooks.com**

ELEMENTS

OF

COMPARATIVE ZOOLOGY

BY

J. S. KINGSLEY, S D.

Professor of Zoology in Tufts College

NEW YORK

HENRY HOLT AND COMPANY

1897.

PREFACE.

THE present volume is intended as an introduction to the serious study of zoology. It embraces directions for laboratory work upon a selected series of animal types and a general account of related forms. Laboratory guides are somewhat numerous, but general outlines of zoology adapted to beginners are few. By combining the two, it has been possible to emphasize the comparative side of the subject. A knowledge of isolated facts, no matter how extensive, is of little value in education, excepting as the powers of observation are trained in ascertaining those facts. Nature studies are truly educational only when the student is trained to correlate and classify facts. A considerable experience with students of different ages has resulted in the conviction that it is not sufficient to ask one to compare a grasshopper and a beetle, pointing out their resemblances and points of difference; leading questions must be asked. When the student has answered the questions under the headings "Comparisons" in the following pages, he has a tolerably complete statement of the principal characters of the larger groups of the animal kingdom.

Several considerations have had weight in the selection of types to be studied in detail. In the first place, so far as possible, these should be such as are readily obtainable in any locality. But there are certain important groups, all the members of which are marine. The forms of these which have been used can be purchased of dealers in laboratory supplies (see Introduction) at a cost of less than sixty cents per pupil. In the second place, the number of forms studied and the extent to which details of structure are worked out must be such that the work outlined can be done by students of average ability, in the time usually allotted to such work in the ordinary course. Especial care has been taken that time shall not be wasted in working out features of no morphological importance. Counting tail-feathers or fin-rays has no place in elementary zoology.

Again, the work has been made largely macroscopic in character. Microscopes are expensive, and many institutions feel that they cannot afford to provide each student with one of these instruments. Then, too, there are enough important facts to be discovered with scalpel and hand-lens. Too many beginners have been lost among cell-theories and drowned in staining-fluids. These properly come after the elements of the study have been mastered.

In order of treatment the author has followed the sequence which he believes productive of the best results. A strictly logical course would lead from the simple to the complex, but in practice this has not been found as valuable as the order adopted here.

A number of illustrations have been prepared especially for this work. Most of the others are credited to the author from which they are taken. It may interest some to know that Figures 2 and 127 were engraved for the second part of Agassiz and Gould's " Principles of Zoology," which was never published.

Tufts College, Mass., June 14, 1897.

CONTENTS.

ELEMENTS OF COMPARATIVE ZOOLOGY.

INTRODUCTION.

Every true teacher must have his own methods, but some suggestions as to the way in which this book is intended to be used may be of value. In the first place, the laboratory work is regarded as most important, since through it the student is trained in observation—a training utterly lacking in all the non-scientific studies of the school curriculum; and also since by it he acquires an autoptic knowledge of the animals studied. It is believed that every point mentioned in the laboratory directions can be made out by students in the high-school grades.

Each student should make all the drawings called for.* Drawing the object seen is one of the greatest aids to observation, and every pupil, no matter how lacking in artistic ability, can make intelligent sketches of all points called for. These sketches have great value for the teacher, since by their aid one can see at a glance any errors or difficulties. All questions asked should be answered in the note-book.

At various points are questions grouped under the heading "Comparisons." These questions are based upon the

---

* The expression " × 2," " × 6," etc., means magnified two times, six times, etc.

1

previous dissections, and are intended to bring out clearly in the student's mind the essential points of resemblance and of difference in the forms studied, and the bearings of the facts discovered. Laboratory work trains the powers of observation; the answering of the questions leads to a systematization of knowledge and an exercise of the reasoning powers. The value of nature studies lies more in the training of the mind than in the acquisition of facts. Hence each pupil should be required to hand in answers to these questions, and to make these answers as detailed as possible.

Following each laboratory section is a general account of allied forms and a statement of the principal characteristics of the group, thus giving a completeness to the knowledge which otherwise would be utterly lacking. In these general statements there are frequent references to the sections where the student has worked out the point for himself. The work throughout is based upon the inductive method, and finally the animal kingdom is shown as a whole.

APPARATUS NECESSARY.

The room used for laboratory purposes should be well lighted and should be furnished with running water. There should be receptacles for waste, and the students should be made to keep everything clean.

The tables for laboratory work should be low (not over 29 inches from the floor), and should afford each student at least six square feet of surface. It is best that there should be no varnish upon them, as this makes trouble when alcohol is spilled.

Each student should have the following instruments: A scalpel; a pair of scissors; a pair of forceps; two dissecting-needles (made by inserting the eye end of a needle in a stick about the size of a lead-pencil); a magnifying-glass (a simple lens of about one-inch focus); a dissecting-pan; a jar of alcohol (70%); a note-book, pencils, and drawing-paper. As the animals to be dissected are small, the instruments should be of moderate size, delicacy being preferable to strength. The dissecting-pans (preferably of copper) should be about 6 by 12 inches, with flaring sides an inch and a half in height. The bottom should be covered to about one quarter of an inch in depth with wax, so that the specimen may be pinned out during dissection. For most purposes it is better if the wax be blackened by lampblack.* At the close of each dissecting period the specimen should be placed in the jar of alcohol for preservation until the next time. For this purpose the three-pound glass butter-jars with screw-tops are good.

The pencils should be hard (6H, Faber), and the points should be kept sharp with a file or emery-paper. For drawings a smooth, hard-surfaced unruled paper is best, Bristol-board, aside from expense, being preferable. The drawings should be in outline only; *shading should not be attempted.* Frequently the use of colored pencils will make the sketches more intelligible, and for this purpose the following conventional colors may be suggested:

Arterial circulation, red. Venous circulation, blue.
Alimentary canal, brown. Liver, green.
Kidneys, purple. Reproductive organs, yellow.
Nerves, gray.

The laboratory should be provided with an oil-stone for

* Instead of wax, the cheaper ozokerite (to be obtained of wholesale druggists) may be used.

sharpening instruments; a pair of bone forceps * for cutting hard substances; a hypodermic syringe and other apparatus for injection (see Appendix); a skeleton of at least one representative of each great group of Vertebrates; and at least one *good* compound microscope.

MATERIALS FOR DISSECTION.

The forms selected for study are, so far as possible, such as can readily be obtained in any locality by taking a little pains at the proper season. There are, however, certain groups of animals which occur only in the sea, and representatives of these must be obtained from the shore. These marine forms selected are Embryo Dogfish (*Squalus*), Squid (*Loligo*), Sea-urchin (*Arbacia*), Starfish, Sea-anemone (*Metridium*), Hydroid (*Pennaria*), and Calcareous Sponge (*Grantia*). The series may be obtained from dealers † at a cost not exceeding sixty cents per student. Orders for these should be placed in the early summer, so that no difficulty or delay may occur later. Much of the other material may be obtained when wanted, but such as cannot be had in the colder months—frogs, tadpoles, snakes, turtles, crayfish, insects, earthworms, etc.—should be collected in the summer and preserved in alcohol or formol ‡

---

* In place of the expensive bone forceps of dealers in surgical instruments one can use the oblique-cut pliers to be purchased at any hardware dealer's.

† Supply Department, Marine Biological Laboratory, Wood's Hole, Mass.

Prof. H. W. Conn, Middletown, Conn.

F. W. Walmsley, Bridgeton, N. J.

Leland Stanford University; Stanford University P. O., California. These dealers issue price-lists.

‡ See Appendix for preservative fluids and methods

for use later. Those which require injection should be so prepared before being placed in the preservative fluid.

As far as possible, all dissections should be performed under water. This buoys up the various parts, and makes their shapes and relationships more evident than they otherwise would be.

REFERENCE BOOKS.

In the classroom there should be some works of reference, and the teacher should have and *use* others. As an aid in selection of these works the following remarks may be of value:

There are a number of guides for the dissection of animals. One of the oldest and best of these is the "Practical Biology" of Huxley and Martin (Macmillan & Co.), which deals with both plants and animals in a thorough manner, although but a few forms are included. Of a somewhat similar character is Dodge's "Elementary Practical Biology" (Harper & Brothers), which enters more into the physiological side of the forms studied. Descriptions of more forms will be found in Bumpus' "Invertebrate Zoology" (Holt), Brooks' "Invertebrate Zoology" (Cassino), and Parker's "Zootomy" (Macmillan & Co.), the latter including only vertebrates. The works of Brooks and Parker are illustrated.

For general accounts of the *structure* of animals, giving general statements for all groups, Jackson's edition of Rolleston's "Forms of Animal Life" (Macmillan) and Gegenbaur's "Comparative Anatomy" (out of print; only to be found second-hand) are good. The general structure of invertebrate forms is covered by Lang's "Text-book of Comparative Anatomy" (Macmillan), Shipley's "Invertebrate Zoology" (Macmillan), McMurrich's "Invertebrate

Morphology" (Holt), and Huxley's "Anatomy of the Invertebrates" (Appleton). Of these Lang's work is the most detailed; Huxley's is rather old; Shipley's is the simplest. The structure of. the vertebrates will be found in Wiedersheim's "Comparative Anatomy of the Vertebrates" (Macmillan), Huxley's "Anatomy of the Vertebrates" (Appleton).

The development of animals is discussed in the following works : Balfour's "Treatise on Comparative Embryology" (Macmillan), Korschelt and Heider's "Text-book of Embryology" (Macmillan; one volume published so far), Hertwig's "Text-book of Embryology" (Macmillan), and Minot's "Human Embryology" (Wm. Wood & Co). Balfour's treatise is a standard, but is rather old. Korschelt and Heider deal only with invertebrates ; Hertwig and Minot only with vertebrates.

Good general zoologies are comparatively few. Under this head are here included works which treat of the structure, development, and classification of animals. Possibly the most widely used work is Claus's "Elementary Text-book of Zoology," 2 vols. (Macmillan & Co.), which, however, is largely based upon European forms. The "Riverside Natural History," 6 vols. (Houghton, Mifflin & Co.), is more popular in style, and deals largely with American animals. Somewhat similar English works are the "Cambridge Natural History," 10 vols. (Macmillan & Co.), and the "Royal Natural History " (edited by Lydekker).

The broader and more general biological principles, without reference to classification and description of forms, may be found in Parker's "Elementary Biology" (Macmillan) and Hertwig's "General Principles of Zoology" (Holt).

Besides these there are a number of good works treating of special groups of animals. The student at the seashore of our New England States finds Smith and Verrill's "In-

vertebrates of Vineyard Sound" indispensable. This was
published in the Report of the U. S. Fish Commission for
1871–2, but separate copies may be had from dealers in
scientific books. Emerton's "Life on the Seashore" (Cassino) covers much the same ground, but in a more elementary manner. For the identification of vertebrates Jordan's
"Manual of the Vertebrates" (McClurg) is the standard.
There are two good works upon molluscs, Woodward's
"Manual of the Mollusca" (London) and Tryon's "Structural and Systematic Conchology," 3 vols. (Philadelphia),
both well illustrated. The insects are treated well in Comstock's "Manual of the Study of Insects" (Comstock Pub.
Co., Ithaca, N. Y.) and Hyatt and Arms' "Insecta"
(Heath, Boston). An older work, but still of great value, is
Harris' "Insects Injurious to Vegetation" (Boston).

There are several works dealing with birds. Of these
possibly Coues' "Key to North American Birds" (Estes &
Lauriat) is most widely known. Ridgeway's "Manual of
North American Birds" (Lippincott) is also good, as is
Chamberlain's edition of Nuttall's "Ornithology" (Boston).

There are also several more special works which are of
great value in the laboratory or study-room. Among these
are Huxley's "Crayfish" (Appleton & Co.), Ecker's
"Anatomy of the Frog" (Macmillan), Darwin's "Earthworms and Vegetable Mould" (Appleton), and his "Coral
Reefs." Dana's "Corals and Coral Islands" (Dodd, Mead
& Co.) is a later work. The teacher will find much valuable material in the zoological articles in the Encyclopedia
Britannica, though these are very unequal in treatment.
Some of the best of them have been reprinted in "Zoological Articles" by Lankester and others (A. & C. Black).

A dictionary of scientific terms is frequently asked for.
Any of the more recent unabridged English dictionaries

will contain almost every zoological term one runs across in most books. Several so-called dictionaries of scientific terms have been published, but as yet not a single one of any value has appeared.

The teacher should remember that science is continually growing, and that text-books and manuals grow old. He should therefore have access to some of the scientific journals. Among those most valuable to the teacher of natural history are the *American Naturalist* (Philadelphia) and *Natural Science* (London). *Nature* (London) and *Science* (New York) are weekly publications which include all sciences.

LABORATORY WORK: FISH.

For this purpose any common fish—perch, sucker, pout, etc.—from ten to twelve inches in length will answer. If time permit, it will prove very advantageous to take two different fishes and work out the following points, comparing their resemblances and differences.

I. EXTERNAL CHARACTERISTICS.

TOPOGRAPHY OF BODY. Distinguish in the fish **anterior** and **posterior**, a back (**dorsum**) and a belly (**venter**), and right and left sides. Make out the regions: head, trunk, and tail. Is there a neck? Where is the mouth? the vent?

How many fins can you find? How many are in pairs? How many single? Are any in the median line of the body? Is there a skeleton to the fins? Could you regard a fin as a fold of the skin supported on **soft** or **spiny rays**?

Of the median fins the **caudal** terminates the tail, the **dorsal** is on the back, the **anal** is just behind the vent. Are there two of any of these? Are the upper and lower lobes of the caudal equal (**homocercal**) or unequal (**heterocercal**).

Can the paired fins be compared in position to your own limbs? By feeling, ascertain if there be any solid support in the body for either pair. How does this condition compare with that in man? The anterior paired

9

fins are the **pectorals**; the posterior are the **pelvic** or **ventral fins.**

INTEGUMENT.   On the trunk and tail are **scales.**   Are they regularly arranged ?   Are there scales on the head ? Do they extend on the fins ?   Is there any skin over the scales ?   Is there skin on the head ?   Can you trace the skin of the head into the mouth ?   Find dark **pigment** spots on the body.   Does the color belong to the scale or to what ?   Settle by pulling out a scale.

Notice the **lateral line** running along a row of scales on either side of the body.   Does it continue on the head ? Examine the scales with a hand-lens and see what causes the line.   Examine any scale with the hand-lens.   Is it margin regularly rounded (**cycloid),** or is it toothed or spiny behind (**ctenoid)** ?

THE HEAD.   How many eyes are there ?   Where are they placed ?   Are they movable ?   Are eyelids present ? Notice in each eye the colored **iris** around the central black **pupil.**

What is the position of the mouth ?   See that it has a bony framework, the upper jaw being composed of a **premaxillary** in front, and behind this a **maxillary** which when the mouth is open slides over the **dentary** or lower jaw.   Do any of these bones bear teeth ?   Open the mouth and examine the tongue.   How much can it move ?   Can you find teeth anywhere inside of the mouth ?   Feel with a pin.

How many nostrils, and where situated ?   Probe with a bristle.   Do they communicate with the mouth ?   Can you find any ears ?

THE BRANCHIAL APPARATUS.   Find the **gill-opening,** a crescentic slit on the side bounding the head behind.   In front of it is the **gill-cover** or **operculum,** which may be divided into the operculum proper (composed of several parts) and the **branchiostegal membrane,** supported by the bony **branchiostegal rays,** which completes the apparatus below. Connecting the branchiostegal region with the trunk is the narrow **isthmus,** separating the gill-openings of the two sides.

Lift the operculum and see the **gills.** Each is composed of rows of red **gill-filaments** supported on a **branchial arch.** Between the successive arches are the **gill-clefts.** How many are there of these? Open the mouth and see how the gill-clefts are connected with the posterior part **(pharynx)** of the cavity. Could you regard them as slits in the wall of a tube? Notice that each arch contains a solid support. Can you see a red blood-vessel running along each arch?

Draw a sketch of the left side of the body, inserting and naming all parts that can be seen from the surface.

INTERNAL STRUCTURE.

With scalpel and forceps remove a piece of skin from one side of the fish, exposing the underlying muscles. Notice that these are arranged in chevron-like plates, each plate **(myotome)** extending from back to belly, and being divided into dorsal and ventral portions. Pick among the ventral parts of the muscle-plates. Do you find any ribs? How are they arranged with regard to the myotomes?

Open the fish by cutting with the scissors from just in front of the vent, forward, in the median line, to the pectoral fins, taking care to cut nothing but the body-wall. Make other incisions transverse to the first, so that the body-wall on either side may be turned out like a flap, thus opening up the **body-cavity,** or **cœlom,** containing the viscera. Without further dissection notice the membrane **(peritoneum)** lining the cavity. Is it silvery or pigmented?

In the front part of the cavity is the large reddish or brownish **liver** ; turn this over to the left and expose the **stomach**, connected apparently with the front wall of the body-cavity. Pass a probe from the mouth through the **œsophagus** or gullet into the stomach. From the stomach the **intestine** passes back to the vent. From what part of the stomach does it arise.*   Is it straight? How is it supported in its position ?

Study the liver more carefully. On its anterior surface see blood-vessels (**hepatic veins**). Where do they go ? On its posterior surface is the thin-walled green or yellow **gallbladder**. Can you trace any connection between liver and intestine ?

Where is the thin membrane (**mesentery**) supporting the intestine attached to the body-wall ? Can you find blood-vessels in it ? From where do they seem to come ?

Pull the intestine to one side, and expose the **reproductive organs** in the posterior part of the body-cavity. The **testes** are usually white, the **ovaries** yellow or pink. Both vary in size according to the season. Are either of these structures paired ? Trace their ducts backwards, and see where they empty. In the dorsal part of the body-cavity look for the **air-bladder** (lacking in some fishes). Can you find a duct connecting it with the œsophagus ?

Make a drawing from the side showing the organs studied, and leaving space for additions. Then cut away these parts and find, dorsal to the air-bladder, the long, dark red **kidneys**. Are they enlarged in front (**head-kidneys**) ? Can you trace the kidney duct ?

Continue the median ventral incision forward between

---

* In many fishes worm-like blind tubes (**pyloric cæca**) arise at the junction of stomach and intestine. Their purpose is to increase the surface secreting the digestive fluids.

the pectoral fins nearly to the isthmus, taking care as before not to cut the underlying parts. Cut away the thin partition (**false diaphragm**) just in front of the liver. This will lay open the **pericardial cavity** (part of the cœlom).

In the pericardial cavity lies the **heart**. It consists of a triangular **ventricle** below (in the normal position of the fish) and a more dorsal auricle. In front the ventricle gives off a blood-vessel, which at first has a conical enlargement (**arterial bulb**), and then is continued forward as the **ventral aorta**. Behind the heart is a blood-cavity (**venous sinus**) extending across the body-cavity in front of the false diaphragm. How are the hepatic veins (p. 12) related to this ?

Inject the blood-system by inserting the canula of an injecting syringe (see Appendix) in the arterial bulb and forcing some colored fluid * forward through the ventral aorta. After the injection follow the ventral aorta forward, tracing its branches (**afferent branchial arteries**) into the gill-arches (p. 11). What relations do these branchial arteries and ventral aorta bear to the pharynx ?

Now cut away the floor of the throat and trace in the gill-arches the **efferent branchial arteries** to their union above the gullet in the longitudinal blood-vessel, the **dorsal aorta**. Can you find this aorta in the roof of the peritoneal cavity ? Could the blood-system, so far as you have studied it, be described as two longitudinal vessels lying on either side of the alimentary canal, and connected by a series of paired transverse vessels ? What must be the course of the blood in the different parts of the system ? Draw a diagram illustrating the relations of the circulatory apparatus to the alimentary canal and gill-slits.

* None of the gelatine mixtures answer well here, as the necessary heat weakens the walls of the blood-vessels.

Pick into the side of the tail until the **backbone (vertebral column)** is reached. Take out a small piece of it and clean it by boiling a few minutes. Wash away the flesh, and see that it is made up of a series of bones (**vertebræ**), arranged one after the other. Examine a single vertebra, making out the following parts: (1) **A body** or **centrum**, shaped like an hour-glass and hollow at either end, (**amphicœlous**). Do the hollows of the two ends connect? (2) Arising from the centrum two bony plates (**neural processes**), uniting above into a single **neural spine.** These together form a **neural arch**; so-called, since the great nervous (*neuron,* nerve) structure, the spinal cord, passes through it. (3) On the opposite or ventral side of the centrum a similar **hæmal arch**, composed of **hæmal processes** and **hæmal spine.**

Examine in the same way a vertebra in the trunk region. Can you find the same parts? Do the ribs correspond to neural arches or to hæmal arches, or are they something different from either?

Draw a front view of trunk and caudal vertebræ, naming the parts

In another bit of the back-bone, near the head, see the spinal cord passing through the neural arch. Can you find any nerves given off from it? How are they arranged?

In the tail region see blood-vessels passing in a similar manner through the hæmal arch (*haima,* blood). Pull apart two vertebræ and see what fills the cavities in the ends.

Cut off the head, and after picking away the muscles at the hinder part of the skull above, carefully slice off the top of the skull with a strong knife, taking only thin slices and exercising great care after the cavity of the skull is exposed. Enlarge the opening by picking, and then with great care pull away the loose gray matter which covers the

white or pinkish brain.   When this is exposed make out in
it the following parts, beginning in front:

(1) The **olfactory lobes** tapering in front into the nerves
going to the nasal pits (p. 10).

(2) Two rounded oval masses (**cerebral hemispheres**)
meeting in the middle line in front, and together constituting the **cerebrum**.

(3) The **'twixt-brain**, also two-lobed, but lying at a
lower level.

(4) The large, paired, rounded **optic lobes.**

(5) The unpaired **cerebellum** crowded in between the
optic lobes behind and extending back over the base of —

(6) the **medulla oblongata**, also unpaired, which in
turn tapers into the spinal cord.

Draw the ' brain from above, three times the natural
size, naming the parts.

Cut off the tops of the various regions of the brain.   Do
you find cavities (**ventricles**) in any of them?   Can you
find any nerves going from the brain ?

Boil the head of another fish for a few minutes, and then
pick away the flesh as far as possible with the forceps,
taking care not to pull any of the bones from their proper
positions.   This will expose the skull, composed of numerous bones.   See that these can be grouped in the following
regions :

(1) The opercular apparatus, consisting of the several
bones composing the gill-cover (p. 10).

(2) The facial portion, made up of the jaws and parts
connected with them; numerous small bones around the
eye, etc.   See how the lower jaw is suspended from the
skull.   Does anything like this occur in man ?

(3) The **cranium**, consisting of a number of bones
which form a box to enclose and protect the brain.

Remove the other bones from the cranium and notice the various openings through which nerves and blood-vessels find passage, and especially the large opening (**foramen magnum**) through which the spinal cord passes from the brain to extend along the back.   Can you find a place especially fitted for the junction (**articulation**) of skull and vertebral column ?

On the sides of the hinder parts of the cranium are the thin-walled **ear-capsules.**   Cut into one and open the sac (**vestibule**) of the ear containing a large ear-bone (**otolith**).

LABORATORY WORK: EMBRYO DOGFISH.

How do the fins compare with those of the fish already studied? Have they a supporting skeleton? Answer by pulling off a bit of the skin from the pectoral fin. Is the caudal fin homocercal or heterocercal (p. 9)?

Place a bit of the skin in a drop of glycerine on a slide, and after an hour examine it under the microscope. Notice the scales. How do they differ from those of the bony fish? This kind of scale is called **placoid**.

THE HEAD.—Are the eyes in the same position as in the other fish? Where are the nostrils? Do they communicate with the throat? Where is the mouth? Open it and look for teeth. Do you find them in the same places as in the bony fish? Is there a tongue? Behind each eye is a hole **(spiracle)**. Does it communicate with the mouth? On the sides of the "neck" occur the gill-slits. How many are there? How does this condition compare with that in the bony fish?

Draw the fish from the side.

Internal Structure.

Open, as in the bony fish, by cutting from the vent forward to the pectoral fins. Make cross-cuts and pin out the walls. Can you see the myotomes (p. 11)?

Trace the **alimentary canal.** In the front part of the

17

body-cavity is the two-lobed liver, and between its lobes find the U-shaped stomach. The intestine begins at the end of the U. Cut off a bit of its wall with the scissors and see the **spiral valve** inside. What function can you suggest for it? Is there a mesentery?

Remove the alimentary canal, and on the roof of the abdominal cavity see two long ridges on either side of the mesentery. The outer ones are the **kidneys,** the inner pair, much shorter, the **reproductive organs.**

Cut off the skin between the pectoral fins and clean the muscles from the support of the fins (**pectoral girdle**) which crosses the median line. Is this composed of bone?

Cut through the pectoral girdle and lay open the pericardial cavity. Is the heart like that of the bony fish? In front of the ventricle is an **arterial cone.** How does this differ from the arterial bulb (p. 13)? Trace the ventral aorta into the afferent branchial arteries by carefully picking away the muscles.

Now cut back from either angle of the mouth along the lower margin of the gill-slits, and turn back the lower jaw as a flap. Now the gills can be studied, and in the cut arches the **gill-cartilages** can be seen. How do the gills differ from those of the bony fish?

Slit the skin on the roof of the mouth and carefully remove it with the forceps. This will expose the **efferent branchial arteries,** which can readily be traced to their union into the dorsal aorta.

Cut off the tail, and in the cut surface make out the following points: In place of bony structures in the position of the centrum (p. 14) a gelatinous **notochord** forming the axis of the column, surrounded by a tough **notochordal sheath.** Above and below neural and hæmal arches surrounding spinal cord and blood-vessels. On the sides of the body, just beneath the skin, find the **canal** of the lateral line (p. 10).

Split the skin in the median line on top of the head and pull it off. On its under surface find the branching canals of the lateral line system.

Now carefully slice off the top of the skull, exposing the brain. Enlarge the opening and compare the brain with that of the bony fish. Notice especially the difference in relative size of parts. Draw the whole brain.

Cut away carefully the side walls of the cranium, exposing the nerves coming from the brain. In this process you will lay open the **semicircular canals** of the ear, behind the spiracle, and, deeper down, the vestibule (p. 16). In this last will be found a granular mass. Examine some of it under the microscope in a drop of water and notice the character of the small particles (**otoliths**). The principal nerves that you will find will be the **olfactory,** going to the nose; the **optic,** arising from the lower surface of the brain and going to the eye ; **the trigeminal,** arising from the anterior sides of the medulla and passing forward to supply the "face." Just behind this is the combined **auditory** and **facial** supplying the ear and face, and still farther back the large **vagus,** which goes back to the gills, the lateral line, and the viscera. Trace these nerves as far as possible, and insert them in your sketch of the brain.

Cut off the snout by an incision passing through the nostril, and in the cut surface see the folds of the olfactory membrane.

Have you found bone in any part of the dogfish ?

COMPARISONS.

Divide a page of your note-book by a vertical line; label one column Bony Fish and the other Dogfish, and in each write the answers to the following questions, numbering them as they are here:

(1) What kind of scales? (2) Where is the mouth? (3) What is the shape of the caudal fin? (4) How do the gills differ? (5) Where are the nostrils? (6) What is the character of the hard parts? (7) Is there a spiral valve in the intestine? (8) What parts occur in the heart? (9) Is there a swim-bladder? (10) Is there an operculum?

SELACHII (Sharks and Skates).

These forms, of which the dogfish is an example, are almost all marine. They are sharply marked off from the Teleosts (p. 24) by several important characters. The body is covered with placoid scales, the mouth and nostrils are always on the ventral side of the body, the caudal fin is heterocercal, the gill-slits (usually five in number) open separately to the exterior, the skeleton is cartilaginous, the heart has an arterial cone, and the intestine is provided with a spiral valve. An air-bladder is lacking. There is usually also a spiracle (p. 17). There are two orders of Selachians.

Order I.—Squali (Sharks).

In the sharks the body is more or less cylindrical, and the gill-slits open upon the sides of the neck. About 150 species are known, some, like the dogfish, being small, others reaching an enormous size. Those forms which feed on fish and the like have sharp cutting teeth, and these are arranged in rows, one behind another, so that only one row is in use at a time, the other serving as a reserve supply if one of the front row be lost. In other sharks, which feed on shell-fish, the teeth are flattened plates, the whole forming a mill for crushing the shells. Most of the species are much like the dogfish in their general appearance, but there are strange forms. Thus in the hammer-head sharks the sides of the front of the head are drawn out like a mallet, the eyes being on the outer ends of the lobes. In the sawfishes the snout is drawn out in a long beak, either edge of which is armed with sharp teeth.

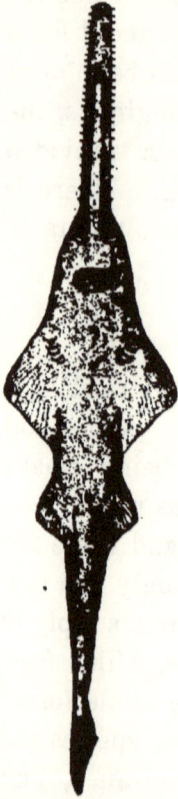

FIG. 1.—Sawfish (*Pristis pecti-natus*). After Goode.

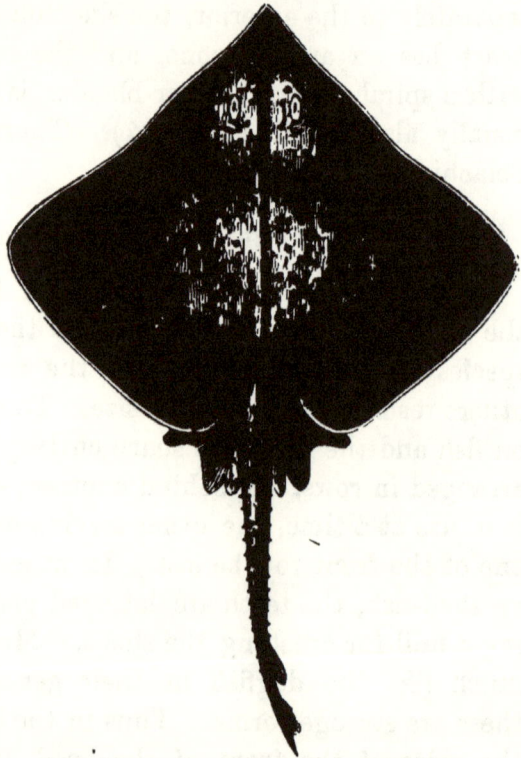

FIG. 2.—Common Skate (*Raia erincea*).

ORDER II.—RAIÆ (Skates, Rays).

In the skates and rays the body is usually flattened, and the gill-slits are on the under surface. In most forms the body is sharply marked off from the tail, but in those saw-fishes which belong to this order the body is shark-like. The width of body in the true skates is partly due to the fact that the pectoral fins are enclosed in it, the whole making a disk, rounded or four-sided in outline. Most of them are bottom feeders, living upon shell-fish, and hence have flattened pavement-teeth. The torpedoes are remarkable for their electrical powers. In them certain muscles on the sides of the head are metamorphosed into an electrical battery, the discharge of which is under control of the will. The current is strong enough to kill small animals which come into contact with the creature. The largest of the skates are the huge tropical devil-fish, which reach a length of twelve or more feet and a weight of 1200 lbs.

TELEOSTS (Bony Fishes).

The great majority of the forms which we ordinarily call fishes belong to the group of Teleosts or bony fishes, so called from the abundant bony matter in the skeleton. In all, the mouth is at the tip of the snout, the nostrils on the upper surface, and the caudal fin, though heterocercal in the young, is homocercal in the adult. The skull is covered with numerous bony plates, and the body is covered with either cycloid or ctenoid scales. Sometimes (trout) scales are apparently lacking, but this apparent absence may be due to their small size and their being buried in the skin. The gills are covered by an operculum. Of the internal features which characterize the group may be mentioned the absence of a spiral valve in the intestine, the presence of an arterial bulb in the heart, and, very frequently, of a swim-bladder.

The thousands of species of bony fishes are variously subdivided by naturalists accordingly as different structures are made the basis of classification. One of the simplest of these schemes recognizes six of these subdivisions or orders, and is adopted here. To which does the specimen you studied belong?

Order I.—Physostomi.

Bony fishes in which the gill-filaments are arranged on the branchial arches like the teeth of a comb; with the pre-maxillary and maxillary bones movable (p. 10); the dorsal, anal, and ventral fins supported only by soft rays (p. 9);

24

and the ventral fins, when present, placed near the vent. A swim-bladder is almost always present, and is connected with the gullet by a tube through which it can be emptied of air. The scales are usually cycloid (10). Most of the forms belong in fresh water.

The catfishes and horned pout, with long filaments or barbels about the mouth, belong here. In our Eastern waters the species are small, but in the Mississippi basin occur large species, some weighing a hundred pounds or more. Many more species occur in the tropics of Africa and South America, and some of these have the scales

FIG. 3.—Atlantic Salmon (*Salmo salar*). After Goode.

developed into a bony armor protecting the body. In Africa occurs a species which, like the electrical eel, can give a severe electrical shock.

The carp and minnows abound in fresh water, but, excepting as they furnish food for other fishes, they are of little importance, the carp of Europe being the least bad food.

Much more valuable is the group of trout and salmon, which are among the most important of food fishes. As a rule these have a soft fin behind the rayed dorsal. The salmon, of which there are one species on the Atlantic and four on the Pacific coast, live in the sea and come into the rivers to lay their eggs. The whitefish of the lakes are closely allied forms.

The blind fish of Mammoth Cave should be mentioned here. In this form a life in total darkness has resulted in the degeneration of the eyes, which are buried beneath the skin.

The savage, swift-swimming pike, pickerel, and muska-longe, the latter reaching a length of eight feet, are, with one exception, confined to America. They are noted for

FIG. 4.—Herring (*Clupea harengus*).

their voracity, and have been termed "mere machines for the assimilation of other organisms."

Among the marine members of the order are the herrings, shad, menhaden, fishes of great importance to man, both as food and for the oil and fertilizers which are made from them. They occur in large schools, and afford food for numerous predaceous fishes.

Differing from the forms already mentioned are those which may be grouped together as eels, fishes with elongate bodies and without ventral fins. Most of the species are marine, and those which live in fresh water go to the sea to spawn. All are voracious creatures, and one South American species has marked electrical powers.

ORDER II.—ANACANTHINI.

These have the gills comb-like (p. 24); the dorsal, anal, and ventral fins without spines; the ventral fins, when

present, placed far forward between the pectorals; the
swim-bladder without connection with the gullet; and the
scales either ctenoid or cycloid. Mostly marine.

FIG. 5.—Cod (*Gadus morrhua*). After Storer.

But few of these forms need mention. Most important
of all are the cod and haddock, which stand beyond all
others as food fishes. They occur in the northern parts of

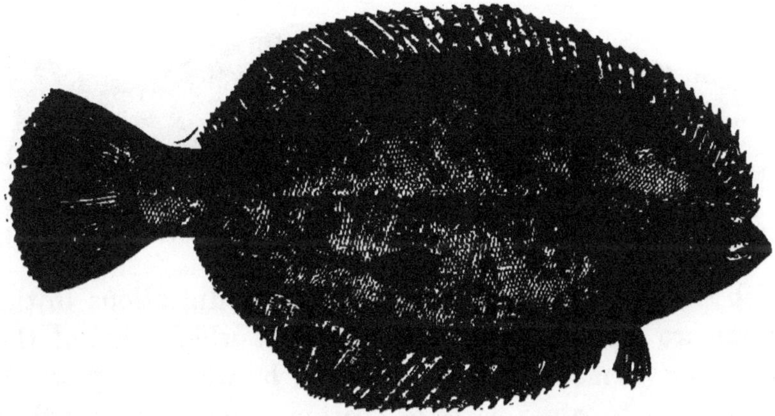

FIG. 6.—Winter Flounder (*Pseudopleuronectes americanus*). After Goode.

both oceans, and find their favorite feeding grounds on
those shallow spots known as "banks." The Grand Banks
of Newfoundland are constantly visited by fishermen from

Europe and America, and have aptly been said to be the richest banks in the world, honoring every draft upon them.

Allied to the cod is the strange group of flat fishes, the halibut, flounders, turbot, and the like. In early life these are symmetrical like other fishes, but as they grow older they turn over on one side, and then the eye of that side migrates to the upper surface, twisting the bones of the skull in its progress. Henceforth the fish lives constantly in this peculiar position, the side of the body turned downward being white, the other colored. The halibut, occurring in all northern seas, are among the largest fishes, occasionally weighing 350 to 400 lbs.

ORDER III.—ACANTHOPTERI (Spiny-finned Fishes).

In this, the largest order of bony fishes, the gills are comb-like, the jaw-bones are movable (p. 10), and the dorsal, anal, and ventral fins have spiny rays in front. In some there is a swim-bladder, but it is without connection

FIG. 7.—Remora (*Remoropsis brachyptera*). After Goode. The sucker is shown on the top of the head.

with the gullet. Among the strange modifications in the group are the suck-fish or Remoras, in which part of the dorsal fin is modified into a sucker, by which they attach themselves to other fishes or floating objects, and are thus carried about.

In the swordfishes the bones of the upper jaw are modified into a long, stiff sword terminating the snout, and used as a weapon of offence and defence. The largest species

reaches a length of fifteen feet. In other points of structure
the swordfish are much like the mackerels, pompanos, and
bluefish, so well known as food fish. Of these the largest

FIG. 8.—Mackerel (*Scomber scombrus*).

is the tunny or horse-mackerel, which sometimes weighs
1500 lbs.

In another group of perch-like forms the spines of the
fins are more developed. Here belong the perch, sea-bass
and porgies, the sheepshead and sunfish, sculpins, and a
long series too numerous to mention.

ORDER IV.—PHARYNGOGNATHI.

These are Acanthopteri in which the last branchial arches
are fused into a single bone, which thus resembles an addi-

FIG. 9.—Cunner (*Ctenolabrus cœruleus*). After Goode.

tional jaw in the throat, whence the name. All of the
species are marine, and with few exceptions they are trop-

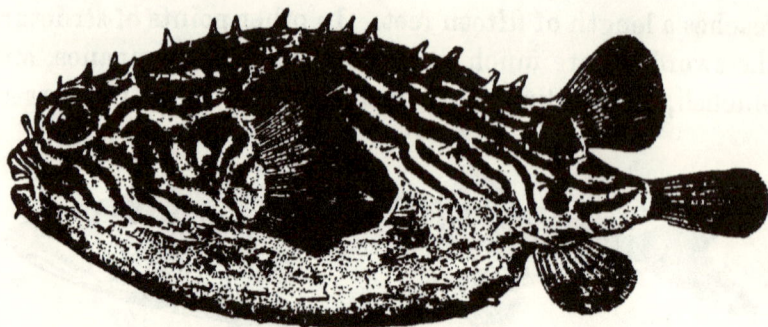

FIG. 10.—Swellfish (*Chilomycterus geometricus*). After Goode.

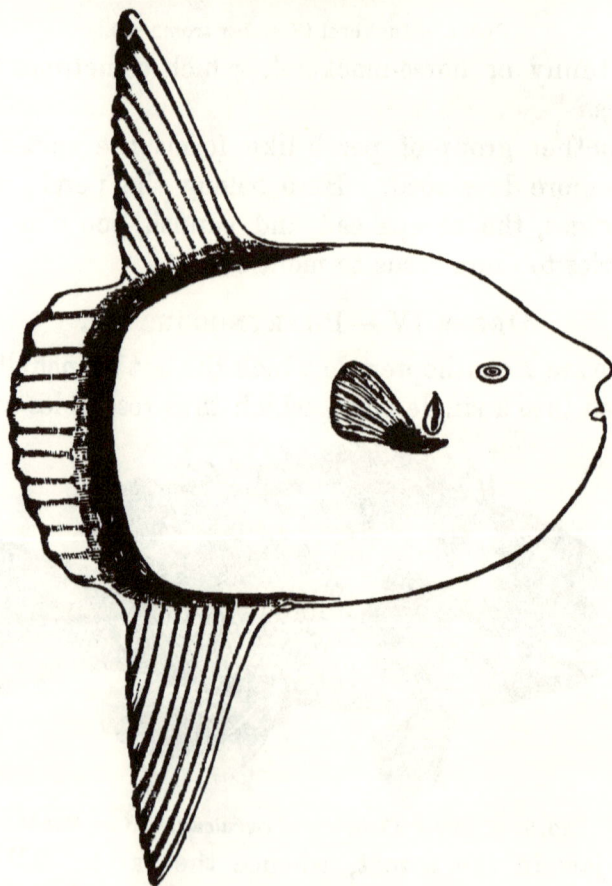

FIG. 11.—Sunfish (*Mola rotunda*). After Putnam.

ical. On our east coast are found the cunner and tautog; on the Pacific occurs a group of surf-fishes (Embiotocidæ), remarkable for bringing forth living young.

ORDER V.—PLECTOGNATHI.

In this group of peculiar forms, almost all of which are marine, the upper jaws are immovably united to the skull.

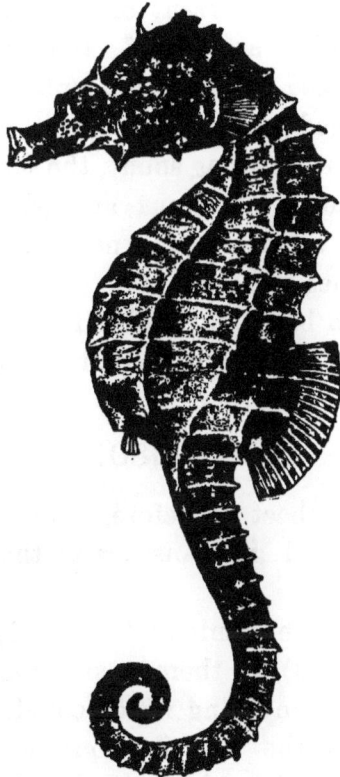

FIG. 12.—Sea-horse (*Hippocampus heptagonus*). After Goode.

Some are naked, others have the skin covered with spines or bony plates. The spiny forms (swellfish) can erect the spines by swelling out the body, and thus gain additional protection. In the trunk-fishes the bony plates unite to

form a solid box. In the sunfishes, which may weigh
500 lbs., the body is almost circular in outline, and has a
distinctly chopped-off appearance. As a whole, the order
bears most resemblance to the Acanthopteri. None are of
the slightest economic importance.

ORDER VI.—LOPHOBRANCHII.

These are the most aberrant of bony fishes. The gills,
as the name implies, are tufted, and composed of small
rounded lobes packed in the gill-chamber. The opercular
apparatus is reduced to a simple plate, the small, toothless
mouth is at the end of a long snout, the skin is covered with
bony plates arranged in rings around the body. The
species, which are all small, are known, from their fanciful
shapes, as pipefishes and sea-horses. Many have a remarkable peculiarity in breeding habits, in that the young are
carried for a time in a pouch beneath the tail of the male.

COMPARISONS.

Prepare another sheet as before, with columns for bony
fish and dogfish, and give answers to the following questions:

(1) Where does the animal live? (2) Is the surface
naked or scaly? (3) Is there a skeleton to the median
fins? (4) Is there anything which could be called a hand
or foot? (5) Do the nostrils connect with mouth or
throat? (6) How does the animal breathe? (7) How
many auricles and ventricles to the heart?

PISCES (FISHES).

The forms to which the name Fishes is usually applied have a body adapted in shape and structure for an aquatic life. It is usually covered with scales, which lie between the two layers (**corium** and **epidermis**) of the skin, the latter extending over them. These scales may be of four kinds, the placoid, ctenoid, and cycloid already mentioned (pp. 10 and 17), and the **ganoid**, either rhomboid or circular in outline, and covered externally with a peculiar enamel layer.

The fins are adapted to fanning the water, being broad plates with an internal stiffening skeleton. Usually both anterior and posterior paired fins are present, and these are supported on skeletal arches or **girdles** (pectoral in front, pelvic behind), which extend around the body beneath, but which have no connection with the vertebral column, nor with any structure like a breast-bone. The pectoral, however, is frequently joined to the skull. The paired fins are largely organs of balancing; the caudal is the chief swimming organ. The caudal fin presents three interesting conditions. In all fishes it is at first **diphy-cercal**; that is, the vertebral column runs out in a straight line, dividing the fin into equal and symmetrical lobes. This condition is retained in a few forms. In others, with growth, the vertebral axis becomes bent upwards, and a secondary lower lobe is developed which, as it is smaller than the other, gives the heterocercal condition (p. 9).

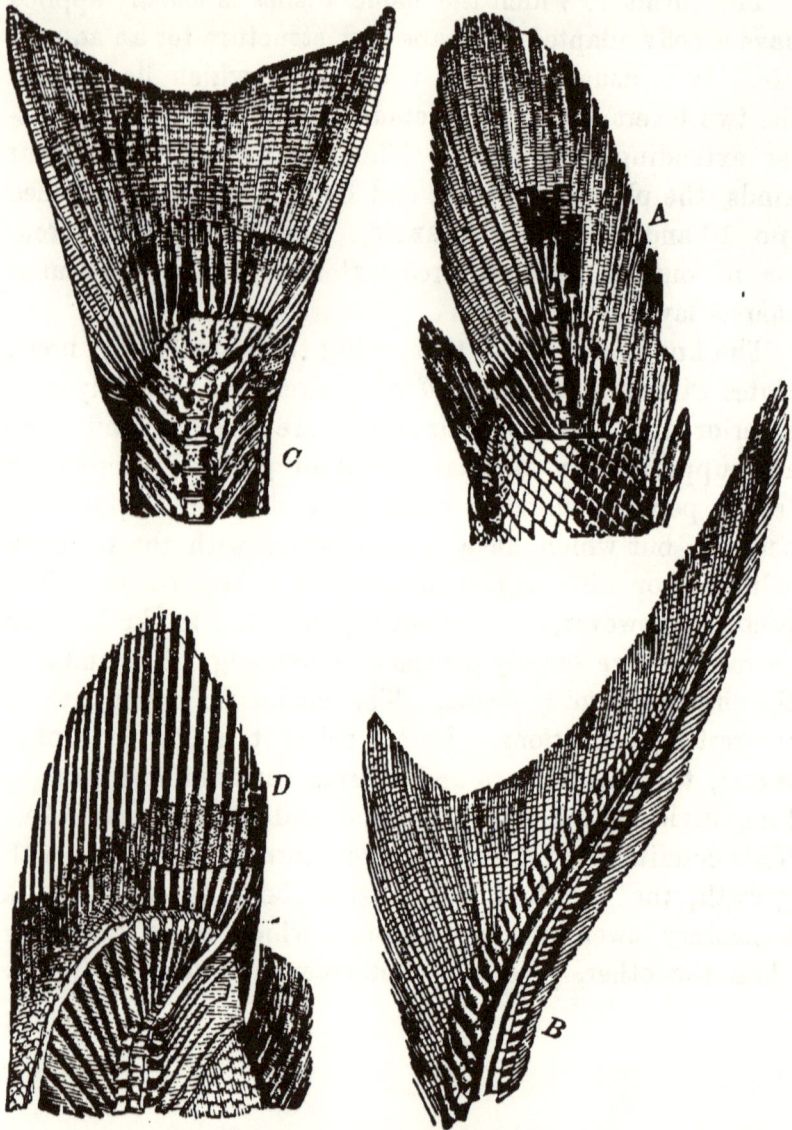

FIG. 13.—Different Forms of Tails of Fishes. *A*, diphycercal; *B*, heterocercal; *C*, *D*, homocercal.

This condition is permanent in the Selachii and most ganoids, but in the bony fishes the lower lobe grows out equal to the other (the tail becomes homocercal (p. 9), although the skeleton shows a bent back-bone (Fig. 13, *C D*)).

The nasal sacs are two, although occasionally four nostrils are present. In no case is there a passage through them to the mouth-cavity, although it is interesting 'to note that in the skates a groove leads back from each of these organs to the mouth, recalling a transitory condition in the young of higher forms.

The gill-slits start as paired outpushings from the throat, which later break through to the exterior. These may all retain their separate external openings, or they may be covered up by a fold from the back side of the head growing over them and forming an operculum. Water taken in through the mouth is forced out through these slits, and is thus brought in close contact with the thin-walled gills lining their sides.

In many forms an air-bladder occurs. This starts as an outgrowth from the *dorsal* wall of the œsophagus or gullet, and in many this connection persists throughout life (Physostomi), but in others the duct is closed later. The bladder serves as a hydrostatic apparatus, and when it is expanded the specific gravity of the fish is lessened and the animal can rise, while when it is compressed the animal sinks. In some forms the bladder is used in producing a noise.

In all fishes the heart, situated in a pericardial chamber, consists of two portions: an auricle, which receives the blood returning from the body, and a ventricle, which forces it forward through the gills to all parts of the animal. In leaving the heart proper the blood first passes through an arterial cone or an arterial bulb. These differ in this: the arterial cone is really an outgrowth of the heart, and contains, on its interior, valves to prevent the flow of the blood back into the ventricle; the arterial bulb, on the other hand, is merely a muscular thickening of the ventral aorta, and contains no valves.

The blood, returning to the heart, bears with it the waste from all parts of the body, and prominent among these is

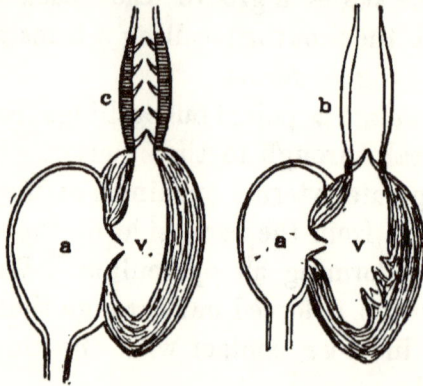

Fig. 14.—Types of Fish-hearts. *a*, auricle: *b*, bulbus; *c*, conus; *v*, ventricle.

carbonic dioxide; in short, it is what physiologists call **venous blood.** This is forced forward, through the ventral aorta and the branchial arteries, to the gills. Through the thin walls of these it comes in close connection with the water, and the carbonic dioxide is given off, while oxygen, from the air dissolved in water, is taken into the blood, which thus becomes **arterial blood,** and is distributed to all parts of the system through the dorsal aorta and other vessels.

It is interesting to note why a fish dies when taken from the water. It is simply because it cannot obtain air enough. When the fish is in the water the gills are floated out so that all parts of them are exposed to the stream passing through the gill-slits. When the fish is out these delicate filaments mat together, reducing the surface for breathing; and then, too, the gills soon become dry, and then are less favorable for the exchange of carbonic dioxide and oxygen.

Among the peculiarities of the skull are the numbers of branchial arches and the ease with which these, the opercular structures, and bones of the face can be separated from the cranium (p. 15). In the Selachii these, like the rest of the skeleton, are composed of cartilage. In the Teleosts this is largely replaced by bone. Another peculiarity is that the lower jaw does not directly join (**articulate** with) the skull, but certain parts intervene between the two, forming what is known as a **suspensory apparatus**.

The group of Pisces is divided into five **subclasses**.

SUBCLASS I.—SELACHII (p. 21).

SUBCLASS II.—HOLOCEPHALI.

A group of less than ten species of strange marine cartilaginous fishes in which the upper jaw is firmly united to the cranium, the gills are covered by an operculum, and a spiracle is lacking. Mouth and nostrils are ventral, as in

FIG. 15.—*Chimæra monstrosa.*

the sharks. The name Chimæra, given to some forms, emphasizes their strange appearance.

SUBCLASS III.—GANOIDS.

These are remnants of a group once very abundant on
the world's surface, but now showing less than fifty living
species in the whole world, and most of these in North
America.   Some of them are much like Selachians, others
like Teleosts, and still others go off towards the Holocephali.
The skeleton is bony or cartilaginous; the body may be
covered with ganoid or cycloid scales, or with bony plates,
or it may be naked; the tail either homo- or heterocercal;
the gills are covered with an operculum.   The heart is
provided with an arterial cone, and the intestine has a spiral
valve.   A swim-bladder occurs, and this has its duct, which,
in one form, empties into the *ventral* side of the œsophagus.
With this confusing mixture of characters it is not strange
that many naturalists have split up the group and distributed its members among the other subclasses.

FIG. 16.—Common Sturgeon (*Acipenser sturio*).   After Goode.

To it belong the sturgeons, the most sharklike of all,
some of which live in fresh water, while the marine forms
ascend the rivers to lay their eggs.   From their ovaries are
made caviare, while their swim-bladders furnish the isinglass,
now so largely supplanted in domestic economy by gelatine.
Though some attain an enormous size, all feed upon small
animals, worms, insect larvæ, etc., which they find in the
mud.   The garpikes, with their strongly armored bodies,
which also belong here, on the other hand, are very vora-

cious. The bowfin of the United States is the most like Teleosts of all.

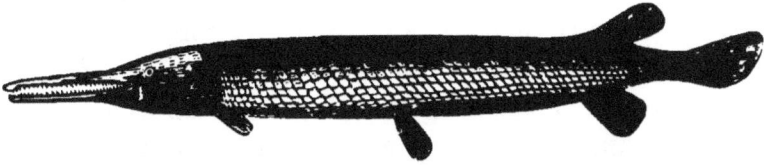

FIG 17.—Garpike (*Lepidosteus osseus*). After Tenney.

SUBCLASS IV.—TELEOSTS (p. 24).

SUBCLASS V.—DIPNOI (Lung-fishes).

Three or four species, one from Australia, one from Africa, and one or two from South America, are the sole living representatives of this group, which however occurs

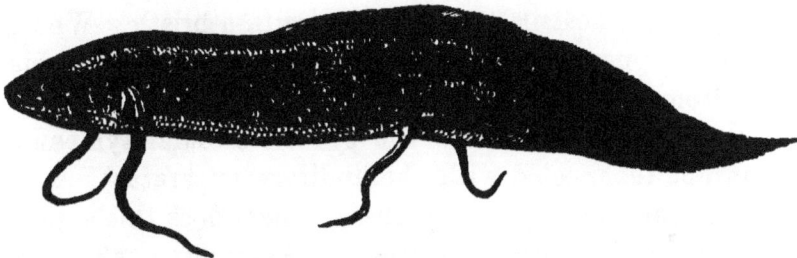

FIG. 18.— Lung-fish (*Protopterus annectens*). After Boas.

as fossils in very old rocks. They have scaly bodies, diphycercal tail, spiral valve, and a swim-bladder which is used as a lung. Both pectoral and ventral fins are present, and these are supported by a peculiar skeleton, while the skull shows many strange features.

LABORATORY WORK: FROG.

If live frogs can readily be had, the student should have a chance to study them alive before dissection. Notice the way in which the eyes can be retracted. Notice especially the way in which the frog breathes. Watch the nostrils during the operation. On the back, a little in front of the vent, may be seen a pulsation. This is produced by **lymph-hearts** beneath the skin. Kill the frogs by wrapping them in a cloth moistened with chloroform, and put them in a close jar for an hour.

Notice the shape of the body. Can you find scales anywhere? Is there anything like a tail? How many appendages are there? How do they compare with your own limbs? Open the mouth; where do you find teeth? Where are the nostrils? Probe them with a bristle. Where does this appear in the mouth? How does the tongue differ from your own?

Behind and a little below the eye is a circular **tympanic membrane** (connected with the auditory apparatus). Cut through this and insert a probe. Where does this appear in the mouth? With what does this **Eustachian tube** most nearly correspond in the shark? See the way the mouth-cavity narrows behind to form the gullet. In front of this see the slit-like **glottis** in the floor of the mouth.

In the fore limbs do you find parts corresponding to arm, forearm, wrist, palm, and fingers? How many fingers?

40

In the hind leg do you find any parts besides thigh, shank, ankle, instep, and toes? If you have any difficulty, compare the way in which the joints bend with those in your own body, and find where your trouble is.

Internal Structure.

Beginning just in front of the vent, slit the skin of the ventral surface in the middle line forward to a point between the shoulders. Turn back the skin on either side. Is it firmly attached to the underlying muscles? Are there blood vessels on its inner surface? Notice the muscles; can you find any muscle-plates (p. 11)?

Next cut the muscles in the same way, a little to the (animal's) left of the middle line, carrying the incision forward through the hard parts between the shoulders, and taking great care to keep the underlying parts uninjured. This lays open the peritoneal cavity (cœlom). Insert a blowpipe into the gullet and inflate the stomach. Is there any sharp boundary between it and the intestine? Is the intestine more or less coiled than in fish or dogfish? Is it of the same size throughout? How is it suspended?

Does the liver cover the stomach? Turn the liver forward and look for the greenish, spherical gall-bladder and the light-colored, lobulated **pancreas**. Do you find ducts from either of these to the intestine? Farther back, in the mesentery, near the enlarged portion (**rectum**) of the intestine, is the red **spleen**. At the posterior portion of the peritoneal cavity is the thin-walled urinary bladder. With what is it connected?

Draw the digestive organs, showing the position of the deeper structures by dotted lines.

Turn the intestines, etc., out of the body, exposing the

reproductive organs and kidneys. These will differ in their appearance in the two sexes.

In the male a yellowish, rounded body (**testis**) occurs on either side of the median line, and just in front of each are the yellowish, lobulated **fat-bodies**. Beneath (dorsal to) the testes are the reddish-brown **kidneys**, each having on its ventral surface a yellowish or golden **adrenal**.

What is the shape of the kidneys? Are the testes and kidneys connected in any way? Do you find the ducts (**ureters**) leading back from the kidneys? Where do they end?

In the female, the **ovaries**, crowded with dark-colored eggs, occur in the place of the testes, their size depending upon the season. Near them are the coiled **oviducts**. Trace these forward and back to their terminations. Do you find the fat-bodies? Do kidneys and adrenals correspond to the conditions described for the male? Are these ureters distinct from the oviducts? Draw the reproductive and urinary organs of your specimen.

Insert a blowpipe in the glottis (p. 40) and inflate the lungs. What is their shape? Are they made up of little chambers (**air-cells**) throughout?

Between lungs and liver is the pericardial cavity, and through its walls in the freshly killed specimen the beating of the heart can be seen. Open the pericardium very carefully and expose the heart; make out the ventricle behind, and the auricles in front. Arising from the ventricle and crossing the auricles is the arterial trunk. Carefully clean this from the surrounding tissues and trace * it to its

* This is best done in an injected specimen. The injection can be made by opening the ventricle and through it inserting the canula into the arterial trunk and tying it there. Then force in the injecting fluid.

division. Then follow each trunk. The right one soon divides into three branches; the anterior is the **carotid**, the middle the **aortic** arch, the third the **pulmonary artery.** How does the trunk of the left side differ?

Trace the carotid arch; where does it go? What becomes of the aortic arch? Do you find a dorsal aorta? On which side of the alimentary tract should the dorsal aorta be (p. 13)? To what organs is the pulmonary artery distributed? Do you find anything to compare with the ventral aorta (p. 13) and afferent and efferent branchial arteries? Draw the circulatory system as made out.

Place a drop of blood of the frog on a slide, cover it with a cover-glass, pressing it well down, and examine under the higher power of a microscope. What is the shape of the **corpuscles?** Are all alike in shape and size? Stain with fuchsin (see Appendix) and study again. Are all parts equally stained?

Split the skin along the back and pull it away. Find the point where the head joins the back-bone; and beginning here, with a strong pair of scissors cut away the roof of the skull bit by bit, taking great care not to injure the brain. Then in the same way cut away the neural arches of the vertebræ. This will expose the brain and spinal cord. The later work will be more easily followed if the animal be put for a day or more in 70% alcohol.

In the spinal cord notice the **spinal nerves** given off at regular intervals on either side. How many are there? What relationship do they bear to the bodies of the vertebræ? Examine these spinal nerves more closely, and see if each is double (has **dorsal** and **ventral roots**). Follow one out by carefully cutting away the bone, and see where the roots unite. Has either root an enlargement **(ganglion)**? Look in the dorsal part of the body-cavity for these spinal

nerves. Trace the posterior ones back to their union
(**plexus**) to form the **sciatic nerve** going to the hind limb.

In the brain, between the eyes, are the cerebral hemispheres. Are they separate? In front are the olfactory
lobes. Are they separate? Behind the cerebrum, and at a
lower level, is the 'twixt-brain. Next come the optic lobes,
and behind them the medulla. What has become of the
cerebellum (p. 15)?

Sketch the brain and spinal cord from above, inserting
all the nerves seen, and making the sketch twice the size of
nature.

Cut across the olfactory nerves and turn the brain backwards. This will show the **optic nerves.** Cut these as far
as possible from the brain, and do the same with other
nerves farther back, at last removing the brain from the
skull.

On its under surface trace the optic nerves back to the
brain. Does the right nerve connect with the right optic
lobe? Behind the optic nerves is a small projection, the
pituitary body. How many nerves can you find arising
from the side of the medulla?

With a sharp scalpel split the brain horizontally and examine the cavities found. Are they all connected? The
larger cavities are called **ventricles.** Those in the hemispheres are the first and second, that in the 'twixt-brain the
third, and that in the medulla the fourth. Are there ventricles in the optic lobes? Draw the brain, showing all cavities and connections found.

From another frog make a skeleton by removing as much
of the flesh as possible with scissors and scalpel, then boil it
with a little soap in the water, and pick away as much
more as you can, taking care not to separate the joints.*

* Much better skeletons can be made by cleaning off the flesh and

In this preparation how many vertebræ do you find? Can you find neural and hæmal arches (p. 14)? On either side of each vertebra find a **transverse process.** How do these compare with the ribs of a fish (p. 14)? Are they the same? Give the reasons for your conclusion. Notice the long bone (**urostyle**) terminating the vertebral column. Connecting the hind limbs with the back-bone is the **pelvic arch.** Is it a true girdle? With what part of the vertebral column does it join? Connected with the fore limbs is the **shoulder-girdle.** Does it join the vertebral column?

Extending along the median line below, in connection with the shoulder-girdle, is the breast-bone, or **sternum.** How many parts in it? Are all equally hard? Connecting the breast-bone with the shoulder are two bones on either side; the anterior is the **clavicle,** the posterior the **coracoid.** Extending dorsally from the shoulder-joint is the shoulder-blade (**scapula**), and above it the **supra-scapula** (partly cartilage). At the junction of coracoid and scapula is the **glenoid fossa,** in which fits the head of the first bone (**humerus**) of the arm. Has a joint like this much freedom of motion? The bone of the forearm is the **radio-ulna.** Does it show any signs of a double condition? With what does it connect below? How many bones in the wrist (**carpus**)? How are they arranged? How many in the palm (**metacarpus**) and in each finger? How does the thumb differ from the others?

On the outside of each half of the pelvic girdle is a deep cup (**acetabulum**), in which is the head of the thigh-bone

then soaking the frog for weeks in water, brushing the parts every few days with a tooth-brush. If such a skeleton be soaked for a few days in Wickersheimer's fluid (see Appendix) and dried, it will retain its flexibility and usefulness for years.

(femur).   Below this comes the **tibio-fibula.**   Is this
double ?   Below this comes the ankle region (tarsus).   The
first two bones of this are long, the second very short.
What effect does this have on the position of the heel (p. 41)?
Compare the tarsus with the carpus.   Is there anything which
you could call a sixth toe ?   Does it come on the inside or
outside of the foot ?

In the skull distinguish between cranium and face (p. 15).
Notice the way in which the upper jaw is attached to the
cranium behind.   Are there teeth on the same bones as in
the teleost ?

THE TADPOLE.

If possible the pupils should have a chance to examine tadpoles of different ages. These can readily be obtained by collecting the eggs in the spring and allowing them to hatch out in glass jars. A number of these can be killed at various stages by means of picrosulphuric acid (see Appendix) used for a couple of hours, then washed two to three hours in water, and preserved in 70% alcohol. The earliest stage necessary should show the external gills, the latest should have the hind legs well formed.

In the earliest of these larvæ the pupil should pay especial attention to the gills; the tail with its fin, how does it differ from that of fishes? In the older larvæ the jaws should be examined. What is their nature? What is the size of the mouth compared with that of the adult? On the left side of the body see the opening of gill-chamber. Is there one on the right side? Carefully open this chamber. Do the right and left sides of the gill-cavity connect? Can you find any traces of the fore limb? Carefully open the abdomen and notice the compact coiling of the intestine. Is it relatively longer or shorter than in the adult? Pick away the muscles from one side of the body until the middle line of the body is reached. Do you find any vertebræ? Lying in this median line find a continuous gelatinous cord, the **notochord**.

COMPARISONS.

Prepare a sheet of paper with two columns as before, one for fishes and the other for the frog and tadpole, and give answers to the following questions:

(1) Is the skin naked or scaly?

(2) What kind of appendages occur?

(3) Is the pelvic girdle united to the back-bone?

(4) Is there an Eustachian tube?

(5) What differences are there in the heart?

(6) What are the organs of respiration?

(7) Do the nostrils communicate with the mouth?

(8) Differences between transverse processes and ribs?

(9) Is a sternum present?

BATRACHIA, OR AMPHIBIA.

The frog may serve as an example of the Batrachia, which, so far as living representatives are concerned, are marked off from the fishes by the features brought out in the comparisons, as well as by a number of other features not easily made out by the beginner. With very few exceptions the Batrachia pass at least a part of their life in the water, and many, in reaching the adult condition, pass through great changes in structure (all are familiar with the change of the tadpole into the frog), so that we must, in considering the group, take into consideration the characters of both **larva** and adult.

In all the skin is very glandular and in all, except the tropical group of blindworms, scales are lacking, and, excepting again these same limbless forms, fins have given place to legs, much like the limbs of man, and like them ending typically with five digits. In the larvæ of all there is a tail, and some (salamanders) retain this structure during life, while in others, as in the frog, it is absorbed (not dropped off) during growth. The larval tail bears a median fin, but this is never divided into dorsal, caudal, and anal (p. 9), and it differs further from those of fishes in having no internal skeleton.

Of internal features those most distinctive are the skeleton of the limbs, unlike that occurring in any fish; the union of the pelvic girdle with the back-bone; the existence of an Eustachian tube in connection with the ear ; the con-

nection of the nostrils with the cavity of the mouth; and
the presence of two auricles in the heart.

In the larva respiration takes place by gills, recalling
those of fishes; and in a few forms these are retained during
life. Besides gills, all, in the adult condition, develop
lungs,* which grow out from the pharynx, and always retain their connection with it by means of a windpipe (trachea) opening upon its floor (compare p. 35). The gills are
fewer in number than in any fish, and only three or four
gill-slits are formed. Between these slits are developed
external gills. Later the slits are closed in most salamanders which lose the gills by the growing together of the
slits. In the frogs the process is preceded by the formation of an opercular fold (compare fishes) in front of the
gill region on either side. These folds grow back over the

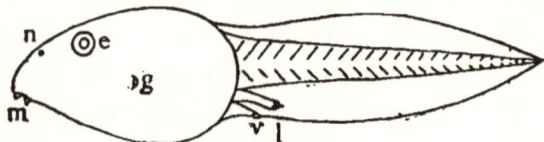

Fɪɢ. 19.—Side View of Tadpole. *e,* eye: *g* gill-opening; *l,* hind leg·
m, mouth; *n,* nostril: *v,* vent.

gill-slits, those of the two sides fusing below the throat and
uniting with the wall of the body above and behind the
gills, thus forming a large chamber outside the gills which
is connected with the exterior by a small opening on the
left side,† through which the water used in breathing
passes.

In the larva the heart is two-chambered, and the blood,

* It has recently been shown that some of the North American
salamanders never develop lungs, but respire solely through the skin.

† Right and left openings occur in two tropical toads (Aglossa).
A few forms have a median opening.

passing forward from it, traverses afferent and efferent branchial arteries, as in fishes, and is collected, as in those forms, in a dorsal aorta. With the loss of gills and the development of lungs the gill circulation changes. The first arterial arch becomes converted into the carotid artery, supplying the head; the second, the aortic arch, connects the heart with the dorsal aorta; the third dwindles and usually disappears; while the fourth, the pulmonary artery, carries blood to the lungs and skin. As will be seen, the embryonic circulation is like that of the fishes, but the different condition in the adult is brought about not so much by new formations as by modifications of pre-existing structures.

In the larva the heart pumps only venous blood, as in the fish. With the development of lungs and the division of the single auricle into two, different conditions occur. Blood from the body (venous) is poured into the right auricle, and blood from the lungs (arterial, because in the lungs it comes into contact with the air) into the left. From the auricles the blood goes to the single ventricle, and thence through the arterial trunk to head, body, and lungs. So at first sight it would appear as if all parts must receive a mixture of arterial and venous blood, but this is not exactly the case. By means which cannot be described here the purest arterial blood goes to the head, the next to the aorta, while the venous blood is sent to the lungs.

In the larvæ of the frogs and toads the mouth is small and the horny jaws are adapted to scraping small plants from submerged objects. Correlated with this vegetable food is an extreme length of intestine, it being a noticeable fact that herbivorous animals require a longer digestive tract than carnivorous forms.

In the larvæ there is also a well-developed lateral-line

system, and this persists to some extent in the adult of the aquatic salamanders, though disappearing in all other forms.

The vertebral column varies greatly in length, and in all except the footless forms it can be divided into neck (**cervical**), breast (**thoracic**), sacral, and caudal or tail regions, the **sacral** being that which connects with the pelvic girdle. In some the bodies of the vertebræ are amphicœlous (p. 14); in most salamanders they are **opisthocœlous** (rounded in front, hollow behind), while in the frogs and toads they are **procœlous** (hollow in front). The transverse processes of the vertebræ are different from anything in fishes in that they arise from the neural arch and not from the centrum. In some forms the ends of these processes are jointed, and from this and other facts they must be regarded as in part equivalent to ribs. It is to be noticed that these ribs never reach the sternum (p. 45), which, by the way, is a structure lacking in all fishes.

A noticeable feature in the Batrachia is the **metamorphosis** during growth, the chief features of which have already been mentioned, the result being that the adult differs very considerably from the young.

All living Amphibia live either in fresh water or on the land ; none occur in salt water. The existing forms are comparatively small, the largest being the giant salamander of Japan, which may be three to four feet in length. Existing Batrachia are conveniently divided into three groups or orders.

ORDER I.—URODELA (Salamanders, etc.).

These forms retain the tail throughout life, and have the extremities weakly developed, fitted for creeping rather than jumping. Some live in the water throughout life, while others, as adults, are to be sought in moist places. In some forms the external gills are retained permanently. The order belongs almost exclusively to the northern hemisphere, and is especially well developed in America. Allied to these

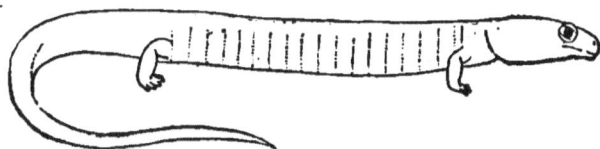

Fig. 20.—Salamander (*Plethodon*).

forms are some enormous fossils, grouped under the name STEGOCEPHALI, some of which had skulls five feet or more in length.

ORDER II.—ANURA (Frogs and Toads).

These in the adult condition lack a tail, and have appendages fitted for leaping. The lower jaw is without teeth. The larvæ are always tailed, and have at first external gills. Frogs and toads differ in that frogs have a smooth skin, and teeth in the upper jaw; toads have a warty skin (caused by numerous glands) and no teeth. Tree-toads are more frog-like, but they have sucking disks on the ends of the toes, by means of which they are adapted to a life in trees. Another group occurs in the tropics, in which the tongue is absent.

Some of the Anura have strange breeding habits. Thus in the European *Alytes* the male wraps the long string of eggs about his body and carries them there until they hatch. In *Nototrema* of South America the skin of the back forms a pouch, in which the eggs are carried; while in the Surinam toad (*Pipa*) the skin of the back becomes very much thickened, leaving little cups, in each of which an egg occurs, and here the young are hatched out.

Another interesting form is the flying tree-toad of the East Indies, in which the feet with the web between the toes become greatly enlarged, forming large disks, upon which the animal sails, much as does a flying squirrel upon its lateral folds of skin.

ORDER III.—CÆCILIA (Blindworms).

These are legless, worm-like Batrachia found in the tropics of both hemispheres. They have a rudimentary tail, degenerate eyes, and the larvæ, so far as known, have three pairs of gills. Some species form an exception to all living Batrachia in having scales in the skin.

COMPARISONS.

Fishes and Batrachia. For the latter use your notes and the preceding account.

(1) Is the blood cold or warm ?

(2) Are median fins present ?

(3) Are gills present in young or adult ?

(4) Are lateral-line organs present ?

ICHTHYOPSIDA (Fish-like Forms).

Under this name are grouped fishes and batrachians, since they are alike in certain important respects. Thus they have, either as larvæ or adults, functional gills, they have lateral-line organs, they have median fins, and the blood is cold. Besides these there are several other points of union, notably in the development, especially prominent being the absence of two embryonic structures, the **amnion** and **allantois**, which occur in higher forms. The Ichthyopsida is divided into two classes:

CLASS I.—PISCES (FISHES) (p. 33).
CLASS II.—BATRACHIA, OR AMPHIBIA (p. 49).

55

LABORATORY WORK: TURTLE.

EXTERNAL.

The hard shell is composed of a dorsal portion, the **carapace,** and a flat ventral shield, or **plastron.** Are the plates covering these arranged in the same way on both ? How are carapace and plastron united ? Are head, legs, and tail naked ? How many toes on the feet ? Are claws present ? Open the mouth. Are teeth present ? Are there lips ? Is there a tongue ? Do the nostrils connect with the mouth ? At the inner angle of the eye see a fold, the **nictitating membrane.** Pull it out with the forceps. What purpose can it fulfil ? Is there an external ear ?

INTERNAL.

Open the body by sawing the hard parts connecting carapace and plastron on either side, then cut the skin, etc., from the plastron, and remove that plate, leaving the animal in the carapace. This exposes the muscles and the limb-girdles, and, after the removal of a thin membrane, the viscera. Was either girdle fastened to plastron ? Just behind the shoulder-girdle is the heart, and on either side of this the dark liver. In the left lobes of the liver is the stomach. Trace the intestine to the vent. Is there an enlarged terminal portion ? Is the intestine supported by a mesentery ? Do you find pancreas or spleen ? Turn the liver inwards and see the lungs. Are they large ?

In the heart how many chambers ? From the front see
56

the vessels. Trace them out, making out carotids, aortic arches, and pulmonary arteries, comparing your work step by step with the frog. (The third of the primary arches has entirely disappeared.) What differences do you find between right and left aortic arches?

In the body-cavity, behind, are the kidneys. Are they smooth or lobed? Where do their ducts empty? Do you find a **urinary bladder** arising from the intestine behind? The ovaries are a broad oval, and can usually be recognized by the contained eggs. Where do the **oviducts** empty? The testes are smaller, long oval, and are outside and behind the kidneys.

In the skeleton * look for the vertebral column on the inside of the carapace. Is it firmly united to it? Can you find any traces of ribs? If so, in what respects are they peculiar? What parts can you recognize in the shoulder and pelvic girdles? In either limb, beyond the humerus or femur, make out two bones (**radius** and **ulna** in the fore limb, **tibia** and **fibula** in the hind limb), and beyond this the (how many?) carpal or tarsal bones. How does this explain certain peculiarities in the frog? Draw either limb, naming parts, remembering that the radius is on the side of the thumb, the tibia on that of the big toe.

In the skull is the socket (**orbit**) of the eye completely enclosed in bone? How does the lower jaw join the skull? What is the means for articulation of the skull with the vertebræ of the neck? Are the vertebræ of the neck hollow in front, behind, or on both surfaces? What name is to be given to the condition found (p. 52)?

---

* Skeletons sufficient for these purposes can readily be made by boiling the specimens and washing away the flesh, with the aid of a nail-brush. It is well to boil the head separately.

Are there any traces of limbs? Can you divide the body into head, neck, thorax, and tail? If so, give reasons for the divisions you recognize. What is the character of the skin? What marked difference exists between the skin of the head and that of the body? Are the dorsal and ventral surfaces alike? Where is the vent? Examine a scale carefully. Is there any skin outside it? Can you pull the scales away from the body? Does a snake shed its skin?

Examine the head. Open the mouth. Are teeth present, and, if so, where? See the tongue. What is its character? Pull it out with the forceps.

58

The following account will apply to almost any common bird. The English sparrow or the pigeon is possibly the most convenient.

EXTERNAL.

Notice that the body presents the regions, head, neck, trunk, and tail. How many paired appendages are found? What covers the body? what the legs and feet?

In the head notice the beak, composed of upper and lower **mandibles**. With what is it covered? Is the upper mandible movable? Open the mouth; do you find teeth? What is the shape of the tongue? Where are the nostrils? Do they connect with the mouth? Behind the tongue, on the floor of the mouth, will be found the glottis (p. 40). How many eyelids do you find? Look at the inner corner of the eye for the **nictitating membrane**. Pull it out with the forceps. Is it like the same structure in the turtle? Hunt among the feathers for the ear-opening. Are the feathers around it different from the others?

Extend the wing. Can you find parts corresponding to arm, forearm, and hand? Are the feathers alike in all parts?* How much is the surface of the wing increased by the feathers?

* The feathers on different parts of the wing have special names. The long quills on the hand are **primaries**; on the forearm, **secondaries**; and those on the arm, when they occur, are **tertiaries**. The short

Are the feathers essentially alike on all parts of the body? Are all parts equally well covered? Pull out a large wing-feather and notice the central axis or **shaft** supporting the expanded portion or **vane** made up of small side-branches (**barbs**), and these in turn having smaller branches (**barbules**). Pull two of these barbs apart, watching with a lens to see the part played by the barbules. Are the conditions the same at the base of the vane? Can you find a downy feather among the others? Examine it carefully and see how it differs from the quills described. Pick the feathers from a part of the breast and study one of the pin-feathers. What parts occur in it?

Next pick the feathers from the whole bird. This will be more easily done by dipping it in hot water. When picking the feathers notice that they come from pits in the skin. When the bird is picked, look for these pits. Are they equally distributed on all parts of the body, or are they arranged in **feather-tracts**?

In the leg see the thigh and shank (drumstick). Where is the heel? Does the bird walk on the whole foot? Connecting the shank with the toes is the **tarso-metatarsus**. How many toes? Do they all point the same way?

INTERNAL STRUCTURE.

Cut through the skin in the median line below from the neck to the vent, being careful not to injure the deeper structures in the neck. Pull the skin away. Insert a blow-pipe in the mouth and inflate. This will render the

feathers overlapping the large quills above and below are the upper and lower **wing-coverts**. At the bend of the wing, just outside the primary coverts, are short quills borne on the thumb and forming the false wing (**ala spuria**).

œsophagus very evident, and will show a specialized enlargement, the **crop**, if it exists.   In front of the œsophagus is the ringed **trachea** or windpipe, while on either side are veins (**jugulars**) usually gorged with blood.

Cut through the abdominal walls in the median line from the breast-bone to the vent.   Open, and, after inflating as before, notice the **air-sacs**.   How many do you find?

Next remove the limbs from one (the left) side, cutting the muscles away from the keel of the breast-bone.   Then cut through the ribs where they join the breast-bone, and next sever them near the back, removing the walls of the body from one side.   This will expose the reddish-brown **liver**, and, partially covered by it, the muscular stomach or **gizzard**; farther in front and near the back-bone the lungs, and in other parts the coils of the intestine.   After drawing the viscera in position, proceed with the dissection.

Pull the gizzard back, and inflate, this time through the œsophagus in the neck.   Where is the glandular stomach (**proventriculus**)?   Where does the intestine connect with the gizzard?   Is the intestine the same size throughout?   Is a mesentery present?

In front of the liver is the **pericardium**, containing the heart.   Open the pericardium and trace, as far as possible without injection, the blood-vessels going from it.   Make out the carotids, aortic arch, and pulmonary arteries.   How many of each?   Which way (right or left) do they turn?   Cut out the heart, and cut it open horizontally.   How many chambers are found?   Sketch the circulation as far as made out.

In the hinder part of the body-cavity are the dark-colored **kidneys**.   Are they irregular in outline?   In front of them are the reproductive organs.   The testes are whitish and oval; the ovaries in the breeding season are filled with eggs

in various stages of growth.   Can you trace the ducts from
either kidney, or reproductive organs ?   Where do they end ?

Open the skull very carefully, beginning at the top and
working down on the sides.   If the head be cut off and put in
alcohol for twenty-four hours or more, the parts of the brain
will be better made out.   In front are the cerebral hemispheres, with the olfactory lobes showing in front and below.
Inserted in the angle between the two hemispheres is the
cerebellum, and on either side of the latter, and partially
covered by the cerebrum, are the optic lobes.   What has
become of the 'twixt-train ?   Do you find the medulla ?
Are all the parts of the brain smooth ?   Draw the brain
from above and from the side.

The essential features of the skeleton can be made out
from the same specimen after boiling.   The parts necessary are the head, the shoulder-girdle, wing, leg, and a few
of the vertebræ.   What is the shape of the ends of the
vertebræ ?   In the shoulder-girdle what parts can you
recognize ?   What name must be given to the wish-bone, or
furcula ? (compare the frog.)   In the wing humerus, radius,
and ulna are readily made out.   How many carpal bones
do you find ?   In the " hand " how many fingers can you
distinguish ?   Sketch the carpus and " hand," with the
ends of radius and ulna.   In the leg recognize femur
tibia, and fibula.   Where is the heel ?   What must the
bone above the toes be ?

Are the bones distinct in the skull ?   Move the beak
upon the skull.   Where do the bones slide ?   Connecting
the angle of the upper jaw with the skull is the **quadrate**
bone.   Is it movable ?

COMPARISONS.

With two columns, one for bird, the other for turtle and snake, answer the following questions:

(1) Is the blood warm or cold ?

(2) Are feathers present ?

(3) Are there any wings ?

(4) Is there an elongate true tail ?

(5) Are the carpus and tarsus long or short ?

(6) Are air-sacs present ?

(7) How many aortic arches ?

(8) How many ovaries ?

The living reptiles closely simulate the Batrachia, and in fact the frogs, toads, and salamanders are reptiles in popular parlance. The short-bodied turtles are paralleled by the frogs, the lizards by the sala-manders, and the snakes by the blindworms. Yet the differences between the two groups are many and important.

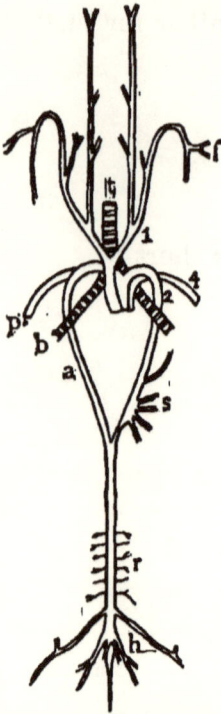

The body is more or less completely covered with scales, and the toes, when present, bear claws. The scales differ from those of fishes in being outside of the outer layer of the skin. These scales differ much in ar-rangement, etc. The large plates covering the carapace of the turtle are but enlarged scales, while the bony armor of the alligator is composed of scales, rendered more protective by the development of bone in the deeper layer of the skin. In the snakes the scaly covering is periodically shed.

FIG. 21.—Arterial Circulation of Turtle. *a*, right aortic arch; *b*, bronchus; *f*, artery to fore limb; *h*, artery to hind limb; *p*, pulmonary artery; *r*, renal arteries; *s*, arteries to stomach; *t*, trachea; 1, 2, 4, persisting aortic arches.

By the greater development of the neck the heart is carried back to a greater distance

64

from the head than in the Batrachia. In all except the
alligators the heart is three-chambered, and in these the
ventricle is incompletely divided into two. There are two
aortic arches, but the left one, which also supplies the
stomach, is smaller where it joins its fellow to form the
dorsal aorta. The blood is cold.

The brain is small, no part being extremely developed,
and the optic lobes touch, or may touch, each other in the

Fig. 22.—Brain of Snake. *c*, cerebrum; *cl*, cerebellum; *o*, optic lobes;
I, olfactory nerve; II, optic nerve.

median line. In snakes, lizards, and turtles the cerebellum is small; in the alligators it is larger.

The vertebræ are usually procœlous, and the vertebral
column is divisible into the regions of neck (ribless),

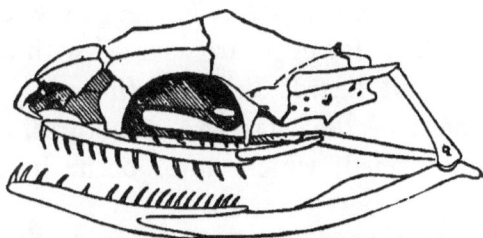

Fig. 23.—Skull of Garter-snake (*Eutœnia sirtalis*), showing the attachment of the lower jaw to the skull by means of the quadrate bone, *q*.
(Slightly enlarged.)

thorax (with ribs), lumbar (ribless), sacrum (usually two
vertebræ which connect with the pelvis), and tail; but in
snakes these distinctions fail, and only trunk and tail vertebræ are recognizable. A breast-bone is present in lizards
and alligators, but none occurs in turtles or snakes. The

skull articulates with the vertebral centrum by a single surface (**condyle**). The hinder angle of the lower jaw is connected with the skull by the **quadrate bone**, which may be free or firmly united to the skull; and the pre-maxillary and maxillary bones are firmly united to the rest of the skull. Teeth are usually present, and in the alligators these are inserted in sockets. The shoulder-girdle (lacking in snakes) is much like that of frogs, the clavicle, however, being absent in alligators. The pelvis is lacking in most snakes, being represented by two bones in the boas. The feet, when present, are usually of the normal type, the bones of the forearm (**ulna** and **radius**) and of the shank (**tibia** and **fibula**) being separate, and the toes, five in number, provided with claws.

In the embryo, gill-slits are partially developed, but no functional gills occur. The lungs are well developed; the left one being reduced or absent in the snakes and snake-like lizards. Respiration is effected by means of the ribs, except in the turtles, and there by a special muscle.

Both ovaries are developed. The eggs are large, and in those reptiles which lay eggs, are covered with a limy shell. A few snakes and lizards bring forth living young. In the development of the eggs two structures, **amnion** and **allantois**, are formed, which never occur in the Ichthy-opsida.

Reptiles are most abundant in the tropics, and are lacking in cold regions. They are mostly flesh-eaters, some living on insects, others on larger forms. Some live on land, some in fresh water, and some in the sea. All living forms can be arranged in four orders.

ORDER I.—LACERTILIA (Lizards).

In these the quadrate bone is movable, but the under jaw cannot be displaced (*cf.* Snakes). Legs are usually present, but either or both pairs may disappear. When the legs are absent the body is exceedingly snake-like, but these forms, like all other lizards, may be distinguished at once from the true snakes by the presence of small scales on the belly. Only one lizard has the reputation of being poisonous, but in former times many, like the basilisk, were fabled to have most deadly powers. Among the more interesting forms are the "glass snakes," so called from the case with which the tail breaks ; the "horned toads," which are not toads, but true lizards; and the chameleons, with their wonderful powers of color change, a capacity

FIG. 24.—Green Lizard (*Anolis*). From Lütken.

which is shared to a less degree by other forms.

ORDER II.—OPHIDIA (Snakes).

These are like the lizards in the movable quadrate, but they differ in the absence of limbs and of sternum, the presence of broad scales (**scutellæ**) on the belly, and in the fact that the lower jaw is connected with the cranium by elastic ligaments, so that it can be displaced in swallowing food.   Many snakes are poisonous, the poison being conveyed into the wound by specialized teeth, the so-called poison-fangs, which are either grooved or are tubular, the grooved teeth being capable of being folded back when not in use, the others being permanently erect.   The rattlesnakes and moccasins belong to the former group.   The largest snakes, the pythons of India and Africa and the boas and anacondas of South America, kill their prey by crushing, as do most of the smaller snakes—our blacksnakes, for example.

Some snakes are protected against their enemies by their colors, which render them inconspicuous in their usual haunts; others by the nauseous smell which they produce by certain glands in the skin; still others by their poison-

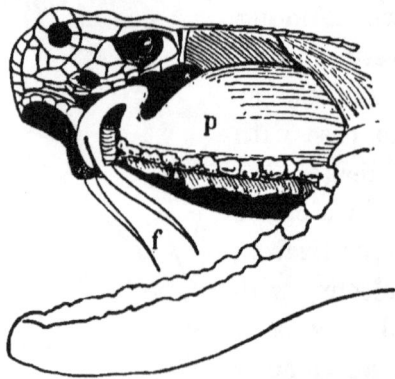

FIG. 25.—Dissection of head of Rattlesnake. *f*, poison-fangs; *p*, poison-sac.

glands.   Most of the snakes are terrestrial, but some, like our water-snakes, take to the water, while in the Indian Ocean are found truly aquatic snakes, which never go on land and which bring forth living young. These sea-snakes are very poisonous. The rattlesnakes are the best known poisonous forms in the United States. In these the rattle is formed by bits of dry skin, which are not lost at the time when the snake sheds the rest of its covering. In this way a new joint is added to the rattle at each molt, and so the whole becomes an approximate index of age.

ORDER III.—TESTUDINATA (Turtles).

The turtles and tortoises are characterized by their short bodies, enclosed in a bony shell or box ; by the absence of teeth ; and by the union of the quadrate bone with the cranium. The shell, with its two parts, carapace and plastron (p. 56), is composed of an outer layer of horny plates (modified scales) and a deeper bony layer, with which ribs and vertebræ are more or less completely united. Into this protective case the head, tail, and legs may be retracted, and in the box-tortoises a hinge in the plastron allows the closure of the openings.

Some turtles are vegetarians, others are carnivorous. Some live on land, some in fresh water, and some in the sea. The largest of existing species are the giant land-tortoises of the Galapagos Islands and Mozambique, and the leather-back and the loggerhead turtles of tropical seas.

Tortoise-shell, before the days of celluloid, was furnished by the dorsal plates of the large tortoise-shell turtle of tropical seas. These plates have the peculiarity that they can be united by heat, so that pieces of any desired size may be obtained. While many turtles are most inoffensive creatures, others, like our snapping-turtles and our soft-shelled turtles, are ferocious, the young snapper showing its temper as soon as it is hatched from the egg.

ORDER IV.—CROCODILIA (Crocodiles and Alligators).

These forms have the highest development of brain and heart of any of the reptiles, the heart being incompletely four-chambered. In general shape they are closely like the lizards, but in bony and other structural features they are greatly different. Crocodiles and alligators are distinguished from each other by the fact that the former have fully webbed feet and more slender snouts. The gavials of the rivers of India have the snout even more slender. The alligators are confined to the New World, while the crocodiles occur in both hemispheres.

The fossil reptiles show a greater range of forms than the living species. The Ichthyosaurs were the whales among the reptiles of former times, while the Plesiosaurs, also swimming forms, had extremely long necks. The Dinosaurs were like the birds in many structural features, although they lacked powers of flight and were terrestrial or aquatic. Some were enormous in size, having thigh-bones nine feet in length and vertebræ five feet across. The Pterodactyls were flying reptiles with wings like those of the bats, except that the wing-membrane was supported by a single finger.

AVES (Birds).

No one can have the slightest question as to whether a certain animal is a bird or not. The feathers, the fore-limbs fitted for flight, and the horny, toothless beak are characteristic of all living forms.

Feathers arise from the outer layer or epidermis of the skin, and each has its tip inserted in a pit or **follicle** in the integument. Feathers vary considerably. Most prominent are the large, strongly built **contour feathers,** which give the animal its general shape. Beneath these are the down and the pin-feathers. Feathers are not uniformly dis-tributed over the body, but are gathered in feather tracts, the arrangement of which varies in different birds. The feathers are not permanent structures, but they are **molted** or shed and replaced by a new growth, this taking place usually once a year. In connection with the feathers should be mentioned the oil-glands (the only glands in the skin of birds) upon the tail, the secretion of which is used in preening the feathers.

In their origin feathers are much like the scales found on the feet, and are doubtless modifications of such struc-tures. The scales on the feet may be small or broad, both kinds sometimes occurring on the same foot. The spur of the cock is but an extremely developed scale with a bony core. These scales differ from those of fishes in that they are developed on the outside of the outer layer of the skin (compare p. 33). The toes are terminated by claws; short

71

in the terrestrial, longer in the arboreal, forms. Claws occur in some cases, especially in young birds, upon the wings.

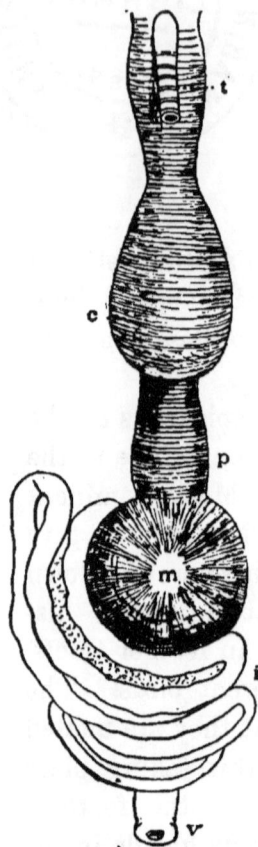

In all living birds teeth are absent, and even in the embryos but the slightest trace of their former existence can be found. In certain fossil birds well-developed teeth occur. The tongue is usually slender, stiff, and horny, and in some forms (woodpeckers, etc.) it is very extensible. The œsophagus is long, and frequently a part of it in the neck is swollen out to form a reservoir of food or **crop**. The stomach is divided into two parts. The first of these **(proventriculus)**, which is glandular, appears much like an enlargement of the gullet. The second or muscular stomach **(gizzard)** is a veritable chewing organ. It is most developed in the seed-eating birds, and in these often contains small stones to assist in grinding the food.

FIG. 26.—**Alimentary Tract of an Eagle.** *c*, crop; *m*, muscular stomach (gizzard); *t*, intestine; *p*, glandular stomach (proventriculus); *t*, trachea; *v*, vent.

The lungs are especially well developed, and a peculiarity is, that connected with them are air-sacs which extend among the other viscera and even into some of the bones, as in the wing.* These air-sacs serve to increase the respiratory surface, and also to lessen the weight of the bird. They are

* A similar **pneumaticity** occurred in the bones of some of the fossil reptiles (Dinosaurs).

also possibly of use in changing the position of the centre
of gravity during flight.

The heart has four chambers, the single ventricle of lower
forms being divided into right and left portions.   The large
blood-vessels which lead from it are, in the embryo, much
like those of the fish; but with development some parts are
altered and others suppressed, so that the result is more modified than in the forms already discussed.   Thus the left half
of the third arch, except for an artery going to the wing of
that side, has entirely disappeared, while the right half, here
called the **arch of the aorta,** connects the left ventricle
with the dorsal aorta.   From this the first arch, modified
into carotids, seems to arise.   The second arch is completely
suppressed, while the fourth arch, arising from the right
ventricle, carries the blood to the lungs.   In returning
from the body the venous blood is emptied into the right
auricle and passes thence, through the right ventricle, to
the lungs for aeration; while that from the lungs goes to
the other side of the heart, and thence to all parts of the
body.   Hence there is here no mixing of arterial and venous
blood in the heart.

In the reproductive organs a constant feature is the
suppression of the right ovary, a rudiment of it existing in
a few forms.   In the breeding-season the oviduct is very
large, and from its walls are secreted the white and the shell
of the egg.   The eggs are large, and are always enclosed in
a limy shell.   There is quite a difference in the condition in
which the young hatch from the egg.   Some are nearly
naked and very helpless, while others are thickly clothed
with down and are able to run and to feed themselves.

The brain is large, and, in comparison with the lower
forms already studied, is noticeable for the great development of the cerebrum and cerebellum, which by their growth

have forced the optic lobes apart and have covered over the
'twixt-brain. The eye is peculiar in that it departs widely from the spherical form, being obtusely conical in front, and in that a circle of bones are developed in this conical portion. There is a tube developed (**external meatus**) leading from the side of the head in to the ear, and this is surrounded by a ring of regularly arranged feathers.

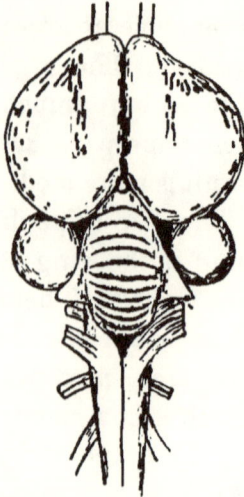

In the skeleton division into neck, thoracic, sacral, and caudal vertebræ, occur. The number of neck vertebræ varies from eight to twenty-four. The sacral are noticeable for their number,

FIG. 27.—Brain of Bird.

and really embrace, besides the true sacrals, some of the lumbars and caudals. The anterior caudal vertebræ are free, but the last six or eight are coalesced into the **pygostyle** or **plowshare bone**. The bodies of the vertebræ in living birds are concave vertically, convex transversely behind, the conditions being reversed on the anterior faces. The cervical vertebræ bear short ribs, free in the young but firmly united in the adult. Each of the true ribs has a small plate (**uncinate process**) on the posterior margin, which connects it with the rib behind. The breastbone (sternum) is large and broad, and in flying birds possesses a strong ridge or **keel** below, to which the muscles of flight are attached. In some flightless birds the keel is lacking.

The skull is noticeable from the great extent of the fusion of the separate bones; for the single condyle for articulation with the neck and for the suspension of the lower jaw by means of a quadrate bone, as in the lizards, snakes, etc.

The shoulder girdle consists of scapula, coracoid, and clavicles, the latter noticeable for their union into a V-shaped " wish-bone " or furcula. In the wing the reduction in bones near the end is remarkable. The bones of the wrist are all united into two, while the three fingers which

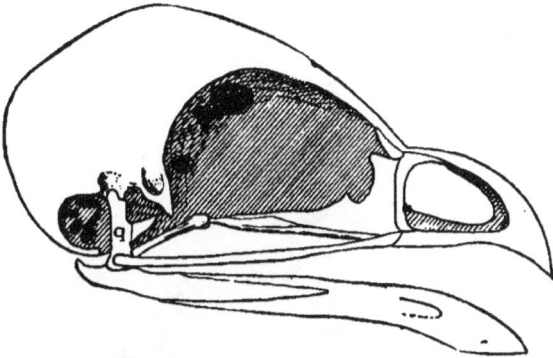

Fig. 28.—Skull of Quail. *q*, quadrate bone.

remain have few joints and are partly united. In the hind limb the fibula is short, but especially noticeable is the great lengthening of two of the ankle-bones, the result being that the heel is elevated some distance from the ground.

Birds are grouped in three divisions or subclasses, the first two of which are extinct; the third contains the ten thousand known species of living forms.

Class I.—Saururæ (Tailed Birds).

These forms, found fossil in the lithographic stone of Bavaria, had tails of extreme length, the feathers being arranged on either side of the long tail vertebræ; and they had teeth in the jaws. Only two specimens are known, the smaller being about the size of a crow, the other somewhat larger.   They are called Archæopteryx.

Subclass II.—Odontornithes (Toothed Birds).

These forms, which have been found only in American rocks, are more like modern birds than is Archæopteryx,

Fig. 29.—Skeleton of Wingless Toothed Bird (*Hesperornis*).   From Marsh.

but they differ from all existing birds in having teeth. They had normal tails, and one form apparently was wingless, only a rudimentary humerus persisting. Some of these toothed birds were about as large as a pigeon; one was about three feet in height.

Subclass III.—Ornithuræ (Modern Birds).

In all living birds teeth are lacking and the tail is reduced; and, excepting a few forms, all have well-developed wings. The recent subdivisions of the subclass are based

Fig. 30.—South American Ostrich or Nandu (*Rhea americana*). From Lütken.

upon characters not readily grasped by elementary students, so we must content ourselves with a classification founded

on external features.   The student should, however, remember that the so-called "orders" are in no wise equivalent to orders in other groups.

Order I.—Struthii (Ostriches).

The ostrich-like birds have long running legs and wings
so reduced as to be useless in flight, and with this the keel
of the sternum (p. 74) has disappeared.   The foot contains
usually three, occasionally but two, toes.   These birds are
mostly large, and embrace the true ostriches of Africa, so
valuable for their feathers; the South American nandus;
the emeus and cassowaries of Australia, and the nearly
wingless kiwi of Australia.

Order II.—Rasores (Scratching Birds).

These, like all the remaining birds, have a keeled sternum.
They have a weakly curved beak, feet well fitted for running, with three toes in front, and a fourth at a higher
level behind.   Here belong the grouse, the pheasants, and
the domestic fowl and turkeys, as well as a considerable
number of tropical forms.   Our common hens, in all their
numberless varieties, are descendants of the wild fowl of
India.   The turkeys are natives of America.

Order III.—Natatores (Swimming Birds).

In these the short feet are adapted for swimming by
having a web between the anterior toes.   The body varies
greatly in shape.   In the penguins the wings have lost the
powers of flight, the wing-feathers being short and scale-
like.   On the other hand, they are strong swimmers, and
the loons almost equal them in this respect.   The other ex-

FIG. 31.—Penguin (*Aptenodytes longirostris*).   From Lütken.

FIG. 32.—Wood-duck (*Aix sponsa*).   After Audubon.

treme is reached in those strong fliers, the albatross, tropic birds, gulls, etc.   More useful to man are the ducks and geese, while the swans, auks, and cormorants must be mentioned as members of the order.

Order IV.—Grallatores (Wading Birds).

The wading birds have long legs, the tarsal region being extremely long, and the shank partly naked.   Correlated with length of leg is length of neck.   Here belong a long series of forms, some of which, like the snipe, are of value

Fig. 33.—Wilson's Snipe (*Gallinago wilsoni*).   After Wilson.

to man as game-birds; while others, like the cranes, herons, storks, etc., have less importance.   Some, like the ibis and the flamingo, are brightly colored, while marabou and egret furnish feathers for human adornment.

In all the foregoing groups of birds the hinder toe is, as a rule, small and of little use.   In all that follow it is usually well developed.

ORDER V.—RAPTORES (Birds of Prey).

The owls, hawks, eagles, and their allies are characterized by short, stout, curved beaks, strong feet and large wings; all structures admirably adapted to the capture of prey and the tearing of flesh. Some, like the eagles, hawks, and vultures, are strong fliers with excellent powers of sight; the owls, on the other hand, are more dependent upon catching their prey by stealth; and their eyes are adapted to their nocturnal habits. The buzzards and vultures depend upon decaying flesh for their food, and their value as scavengers leads to their protection by law in the regions where they occur.

In the birds of prey, like all that have preceded them in our account, the young, when hatched, are covered with feathers (usually down feathers), and have their powers well developed. In all the remaining orders the young are helpless and nearly naked when they escape from the shell.

ORDER VI.—COLUMBINÆ (Pigeons).

The pigeons stand nearest to the Rasores from which, however, they differ in the weaker legs, the large pointed wings, and the fleshy membrane at the base of the beak, pierced for the nostrils. The five hundred different kinds of pigeons show little variety in form. Our domestic pigeons, with their wonderful variations, have descended from the rock-pigeon of Europe. The extinct dodo of the islands east of Africa was a flightless pigeon of large size. The species died out some two hundred years ago.

ORDER VII.—SCANSORES (Climbing Birds).

These birds have the feet adapted for climbing, two of the toes being directed forwards and two backwards. Some, like the toucans, have enormous bills, others have the beak of moderate size.   Here belong the cuckoos, with their reprehensible egg-laying habits, and the well-known wood-

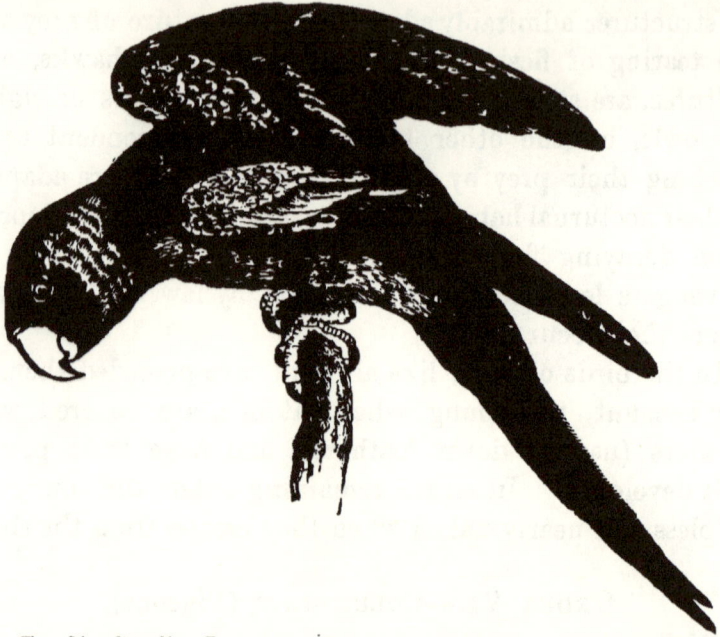

FIG. 34.—Carolina Paroquet (*Conurus carolinensis*).   After Wilson.

peckers.   The large group of parrots also belong to the group of climbing birds.   In these last the tongue is fleshy, and the feet are very efficient organs of prehension.

ORDER VIII.—PASSERES (Perching Birds).

In these the feet have three toes in front, one directed backward and all on a level, and no naked skin on the beak.   They are usually subdivided into the Clamatores or crying birds, and the Oscines or singing birds, the latter having a complicated muscular apparatus in connection with the vocal organs.   To the Clamatores belong the Asiatic hornbills, which recall the American toucans; the

kingfishers, with their large strong beaks; and those gems
of bird-life, the humming-birds. To the Oscines belong an

Fig. 35.—Bird of Paradise (*Paradisea apoda*). After Levaillant.

enormous series of feathered songsters, the mere enumera-
tion of which would take a volume the size of the present

one, the whole series reaching its apex in that pestilential immigrant, the English sparrow. Among these singing birds are some which, like the crows, are not noted for their musical abilities, and their near relatives, the birds

FIG. 36.—Winter Wren. From Coues.

of paradise. We can only mention, in addition, the starlings, flycatchers, wrens, orioles, warblers, and thrushes, forms which make our woods vocal and beautiful.

COMPARISONS.

With columns for Birds and Reptiles, answer the following questions:

(1) Are scales present ?

(2) Are claws present ?

(3) How many occipital condyles ?

(4) Is there a distinct quadrate bone connecting the upper jaw with the skull ?

(5) Are true ribs present ?

(6) How many chambers to the heart ?

(7) What is the size of the eggs ?

(8) Are functional gills ever developed ?

(9) Do the urinary and reproductive ducts empty into the hinder part of the alimentary canal ?

SAUROPSIDA.

Although we naturally associate the birds with the warm-blooded, hair-bearing animals (Mammals), yet structurally they are more allied to the reptiles; a fact indicated by our heading, which means lizard-like. Some of these common features are a body-covering of scales or feathers derived from the epidermis; the articulation of the skull with the neck by a single condyle; the existence of the quadrate as a suspensor (p. 37) of the lower jaw, the presence of true ribs, a three- or four-chambered heart, no functional gills, large eggs with an abundance of yolk, and the existence of a **cloaca** into which digestive, reproductive and excretory organs empty. There are two classes of Sauropsida.

CLASS I.—REPTILIA (p. 64).

CLASS II. AVES (p. 71).

RAT: LABORATORY WORK.

If possible, the student should be provided with two specimens, one injected, the other not. If rats are not easily obtained, a single injected specimen will suffice. Rats may be injected by cutting into the left ventricle and inserting the canula into the aorta from this opening, ligating it by tying in front of the heart. A gelatine mass works best, but care must be taken lest the delicate vessels be broken. Only a small amount of fluid is necessary to fill all the vessels.

EXTERNAL.

With what is the body covered ? Is there hair on the tail ? Do you find scales on the tail ? In what respect do they resemble and in what differ from those of reptile or fish ?

How many toes on the fore-feet ? Do you find any trace of a thumb ? Are the toes provided with claws ? Sketch the sole, bringing out the callous spots. How many toes in the hind-foot ? Sketch the sole and compare with that of fore-foot.

How many nostrils ? Of what use to the animal are the "whiskers" of the upper lip ? Examine eyes and look for third eyelid at inner angle of the eye. Does it resemble any structure you have found in the animals previously studied ? Is there anything similar in your own eye ?

INTERNAL.

Cut the skin along the ventral median line from near the
vent to a point behind the jaw.   Lay the skin back, separating the loose **connective tissue** which binds it to the deeper
parts.   See the thin muscles covering the abdomen.   Feel
for breast-bone, and open up the body by cutting through
muscular walls from between hind-legs to breast-bone.
Make transverse cuts on either side, and fold the walls outwards.   This opens the **peritoneal cavity.**   In this, without disturbing parts, can now be seen, in front, the dark-
colored liver, behind this the coils of the intestine, and between the hinder coils of this tube the urinary bladder.

Tip the liver to your left and find the **stomach.**   Sketch
from the side, showing the entrance of the gullet (**œsophagus**)
and the beginning of the intestine.   Notice how liver and
stomach are connected by thin membrane (**mesentery**).
Tip the stomach forward and notice the **spleen** suspended
in another portion of the mesentery.

Trace the intestine, without cutting anything.   It is also
held by its mesentery.   It makes first a large loop backwards (**duodenum**) and then comes forward to form numerous convolutions.   Find a large pocket (**cæcum**) given off
from the intestine.   All of the tube in front of this is
called the **small intestine**; back of it the **large intestine.**
In the latter two portions—(1) **colon,** (2) **rectum**—can
readily be distinguished by the different appearance of the
walls.

Spread out a portion of the mesentery supporting the intestine and notice in it small vessels.   Some of these will
be found to be single, others double.   The double vessels
are arteries and veins.   They can be distinguished by tracing them towards the middle line of the body.   The veins

unite in a large vein (**mesenterial vein**) which follows along the colon, thence forward into an anterior fold, where it is joined by other veins (**gastric**) from the stomach and (**splenic**) from the spleen. From the union of these is formed the **portal vein** which enters the liver from behind. The small arterial branches arise from a **mesenterial artery** which accompanies the mesenterial vein for some distance and then can be traced back to the median line of the dorsal surface of the body-cavity, where it joins the great arterial trunk, the **aorta**. From the aorta, just in front of the origin of the mesenterial artery, arises the **cœliac artery**, which gives off a branch to the liver (**hepatic artery**), and then divides into **splenic** and **gastric** arteries, going to the spleen and stomach respectively. Trace these arteries. Where does the hepatic enter the liver?

The single vessels in the mesenteries are the **lymphatics**. Their purpose is to carry the products of digestion forward, and eventually empty them into the blood-vessels. These lymphatics unite in a **lymphatic duct**, which runs closely parallel to the mesenterial artery and empties into a **thoracic duct** running parallel with the aorta.

Sketch the blood-vessels (× 2) so far made out, on a sheet large enough to accommodate the whole circulatory apparatus of the rat.

In the mesentery supporting the duodenum find the fatty-looking, irregular **pancreas**. Where does its duct enter the intestine?

How many lobes are there in the **liver**? Are they symmetrically placed? Beside the portal vein and the hepatic artery is the **bile-duct**. Trace it forward and see how its branches arise from the liver-lobes. Trace it backwards and see where it enters the intestine. Look on the posterior surface of liver for the **gall-bladder**. Tip the liver backwards (i.e., towards the tail). See how it is attached by mesenteries to a muscular partition (**diaphragm**) bounding the peritoneal cavity in front. See the œsophagus and a blood-vessel (postcava) extending from the liver through the diaphragm. Sketch the alimentary canal.

Cut through the œsophagus just in front of the stomach and through the rectal portion of the intestine, and cutting the mesentery remove the alimentary canal.

In the body-cavity see, dorsal to the liver, the **kidneys**. Are they at the same level ? Covering the anterior end of each kidney is a triangular **supra-renal capsule**. Trace from each kidney (median surface) backwards a whitish tube, the **ureter**. In the median line of the body-cavity is the aorta already mentioned. Trace it backwards, finding the arteries (**renal**) going to the kidneys. Farther back the aorta divides into a pair of **common iliac arteries**. Trace these into the legs. Do you find them to divide ?

Just behind the point of division of the áorta into the common iliacs can be seen the **common iliac veins**, which return from the legs and unite into a vessel, the **postcava**, which passes forward, at first dorsal to the aorta. A little farther forward the postcava receives an **ileo-lumbar vein** from either side, and then a **renal vein** from each kidney. From the kidneys trace the postcava forward through the liver. This may readily be done by cutting away the ventral part of the liver and then, inserting the point of the scissors into the postcava, make a cut. Continue this until the whole vessel is laid open up to the diaphragm. On the inner surface of the postcava, inside the liver, notice the openings of the **hepatic veins**. These bring to the postcava the blood which entered the liver by the portal vein.

Add these parts to the sketch of the blood system already begun.

With a sharp scalpel split a kidney horizontally. In the cut section make out on the median side a cavity (the **pelvis of the kidney**) from which arises the ureter. Into the pelvis projects from the outer wall a **papilla**. In the thick outer wall notice the difference in appearance between the outer **cortical substance** and the more central **medullary substance**. Sketch the cut section.

Notice the direction of the muscle-fibres in the diaphragm. What would be the effect of their contraction upon the diaphragm? Cut through the diaphragm ventral to the postcava, and continue the cut along the ventral surface of the body to one side of the median line. Cut the ribs with stout scissors. This will lay open the **pleural cavity**.

In the pleural cavity, behind, will be seen the postcava, and dorsal to it the œsophagus. These pass forward between the lobes of the **lungs**. Notice the thin membrane (**mediastinum**) passing dorsally from the breast-bone to the heart and lungs. The heart itself will be found to be enclosed in its own thin sac (**pericardium**). Sketch the contents of pleural cavity.

Cut open the pericardium and study the **heart**. Its **apex** is directed backward and to the (animal's) left ; its broader **base** in front and to the right. Tip the heart to your right, and notice how the postcava enters it near the base on the right side. Just before its entrance into the heart it receives a similar vessel (the **precava**) from in front. Follow the precava forward, cutting away the fatty-looking **thymus gland** just in front of the heart in order to trace the vessel. Soon it divides into right and left branches (**jugulars**) each of which receives a vessel (**subclavian vein**) from the corresponding fore limb. Trace the jugulars forward to head; do they divide ? Insert precava and its branches as well as anterior end of postcava in the sketch of the blood-vessels.

Arising from the left side of the base of the heart is the aorta. Follow this forward; to which side of the body does it turn ? From the arch which the aorta makes trace the following vessels: (1) **Right brachiocephalic artery,** which soon divides into the **right subclavian artery** and the **right common carotid artery.** Follow the subclavian into the limb, and the common carotid towards the head. Where does the common carotid divide into **internal** and **external carotids ?** (Just outside the common carotid will be found a white thread-like nerve. It is the **vagus (pneumogastric) nerve** and supplies the stomach, heart, and lungs). (2) The **left common carotid**; and (3) close to it in its point of origin from the aortic arch the **left subclavian artery.** Trace these as before. Do you notice any differences between these vessels on the two sides of the body ?

Tip the heart to your left and trace the course of the aorta from the origin of the left subclavian back to the origin of the cœliac artery already found. On which side. (dorsal or ventral) of the œsophagus does the aorta pass ? On which side is the heart ? Insert the vessels now made out in the sketch, which should now represent the principal vessels of the **systemic circulation.**

Dissect the aortic arch loose from the surrounding tissue, lift it up, and see dorsal to it the **pulmonary arteries** going to the lungs. From what part of the heart do they arise ? Tip the heart to the animal's right and see the **pulmonary veins,** which bring the blood back from the lungs to the heart. On which side, with reference to the pre- and post-cava, do they enter the heart ? The pulmonary arteries and pulmonary veins belong to the **pulmonary circulation.** Add them to the sketch.

Cut through the cavas, pulmonary vessels, and aorta, and remove the heart. On the base on either side will be found small lobes—the *auricles.* Split the heart with a sharp scalpel parallel to the horizontal plane of the animal, keeping in mind which side of the organ was originally right, and which left. Make out two pairs of cavities (usually containing clotted blood, which should be carefully removed). Which of these has the thicker walls—the right or the left? The basal cavities are the **auricles,** the apical the **ventricles.** Which parts, auricles or ventricles, would you suppose to play the greater part in forcing the blood through the circulation? Study the connections between auricles and ventricles. Do the two auricles connect with each other? Is the same true of the ventricles? Notice what vessels enter the left auricle. Where do the pre- and postcava enter? Where does the blood go from the left ventricle? Insert a diagram of the heart, with its chambers, in the sketch of the circulation.*

Between the common carotids is the ringed **trachea,** or windpipe. Dissect it loose and cut near the head. Insert a blowpipe in the hinder portion and inflate the lungs by blowing. Are the rings of the trachea complete? Trace the trachea forward and notice enlarged anterior portion (**larynx**), and just in front, and ventral to it, the **hyoid bone.** Beneath the trachea (dorsal to it) is the œsophagus.

Remove the skin from the head. Notice the large muscles attached to the jaw, and just in front of the ear the salivary (**parotid**) gland. Cut through the jaw muscles,

* The heart of a cat, sheep, or pig will show these points much better.

and, beginning at the angles of the mouth, carefully cut backwards through the cheeks, so as to allow the lower jaw to be bent back. In the mouth-cavity study the teeth. In front are the incisors, and further back the molars. How many of each in each jaw. With a knife test the hardness of the front and back surfaces of the incisors. Which is the harder? Why are these teeth always sharp? Is there any such arrangement in the molars?

Between the molars is the **hard palate**, its surface with transverse folds. Farther back is the **soft palate**, bounded behind by the place (**internal narial opening**) where the nostrils communicate with the back part of the mouth-cavity. How many of these openings do you find? Slit soft palate with the scissors and see how this arrangement is brought about.

Opposite the internal narial opening (*i.e.*, on the floor of the pharyngeal region) is an opening—the **glottis**, surrounded by a raised rim, which is enlarged in front into a soft **epiglottis**. Inside of the glottis may be seen two folds (**vocal chords**), which narrow the opening. Insert a probe into the glottis. Where does it appear?

Split the skin down the back, and remove it from the body, and then with the bone forceps break through the cranial walls at the back of the head,* taking pains not to injure the underlying structures, When the opening is made enlarge it by removing the skull bit by bit with a strong knife from the dorsal surface and right side. Then continue the process back in the neck region as far as the shoulders.

---

* The points relating to the brain can be made out more easily on the cat or sheep, but with a little pains the directions here given can be followed on the rat.

In the brain make out, viewed from above, in front the **olfactory lobes**; next the large **cerebrum,** and behind this the **cerebellum,** and following the cerebellum the **medulla oblongata,** broad in front and tapering behind into the **spinal cord.** Are any of these parts paired? The line between medulla and spinal cord is not a sharp one, and the place of passage through the skull may be regarded as the boundary. Sketch these parts in outline from above and from the side, ×2.

Over the whole brain is a rather tough membrane, the **dura mater,** which is next to be removed from the dorsal surface. Do you find any convolutions on the cerebrum? Cut through the olfactory lobes as far forward as possible, and lift the cerebrum very carefully from in front. It will be found to be tied by the **optic nerves,** going from the ventral surface. Cut these as close to the skull as possible. Do the olfactory lobes arise from the tip of the cerebrum? Roll the brain very carefully to the left side, looking at the same time at the right side of the medulla for nerves. From its anterior angle (below the cerebellum) will be found a strong nerve, the **trigeminus,** and just behind it another nerve, the **facial** and **auditory** combined. Some distance farther back, yet still inside the skull, arises a more complex nerve, consisting in reality of three, the **glossopharyngeal,** the **vagus,** and the **spinal accessory.** (Thus we can easily make out in the rat the following nerves: I, olfactory; II, optic; V, trigeminal; VII, facial; VIII, auditory; IX, glossopharyngeal; X, vagus or pneumogastric; XI, spinal accessory. The other nerves are not easily made out on so small a form.)

Tip the cerebrum forward, and notice between it and the cerebellum the optic lobes behind and the 'twixt-brain in front. How does this compare with what was found in the dogfish? Tip the cerebellum forward, and see the large triangular opening in the roof of the medulla.

On the lower surface of the brain see the cut optic nerves. From which division of the brain do they arise? Behind the optic nerve find a median lobe, the **hypophysis.**

With a sharp scalpel make a series of cross sections through the cerebrum. Are the two halves completely separate? In each half find a cavity (**ventricle**), and above it in the solid tissue a transverse lighter band (**corpus callosum**). Draw the section. Make similar sections through the 'twixt-brain and the optic lobes. How many cavities do you find here? Draw each section.

Cut a longitudinal vertical section through the cerebellum to left of median line, and notice the way in which the cerebellum is folded. The somewhat bush-like structure is known as the **arbor vitæ.** Make a sketch of it. Cut transversely through the rest of cerebellum and medulla, and in the section see the folds of cerebellum cut in the opposite direction; and, below, the thick floor (**pons varolii**) of the medulla. Cut through the medulla farther back. Do you find a **central canal** in the section?

On the ventral surface of the neck, just outside the carotids, dissect away carefully, keeping the fore legs stretched out, until you find nerves (white cords) going from the vertebral region to the sides of neck. Can you make out two roots to each nerve? Just in front of the ribs notice that the nerves are larger, and that they go to the fore limb just in front of subclavian artery and vein. How many of these nerves as they arise from the neck interlace to form the **brachial plexus** (the network from which the limb nerves arise). Trace them into the limb. Sketch the plexus.

Separate the muscles in the bend of the knee, exposing

the large **sciatic nerve.** Trace the nerve backwards towards the trunk. Does it pass through any bones? Trace the nerve inside the dorsal wall of the body-cavity. Do you find a plexus like that of the fore limb? If so, how many nerves enter into its formation?

COMPARISONS.

With three columns, for Ichthyopsida, Sauropsida, and Rat respectively, answer the following questions:

(1) Is hair present?
(2) Do you find true scales or feathers?
(3) Is there an external ear?
(4) Do you find anything like gill-slits?
(5) How many chambers in the heart?
(6) How many aortic arches?
(7) Do the aortic arches bend to the right or to the left?
(8) Is a diaphragm present?
(9) Do they produce eggs?

MAMMALIA (MAMMALS).

The name Mammalia is applied to all those forms which, like the mouse, cow, and man, have warm blood, a body covered with hair, and which bring forth living young, nourished during the early stages by milk secreted by the mother. These characters at once distinguish any mammal from any other animal, but other features of equal or greater importance occur.

Hair occurs in the young of all mammals, and is usually found also in the adult; but in the case of the whales it is absent in the fully grown animal, and even in the young it is only found near the mouth. Hair is a product of the outer or epidermal layer of the skin. At places this layer dips down into the deeper layer (**dermis**), forming a pit or **follicle** from the bottom of which the hair grows, continual additions being made at this point, commonly known as the " root." The hair itself is a solid column, varying considerably in shape in different animals, from the delicate fur of the fur-seal, to the bristles of the pig or the spines of the porcupine. There are usually glands present which open into the follicle and which secrete a fluid, the object of which is to keep the hair moist; and besides, each follicle is provided with muscles which serve to erect the hair at times of fright (as in cats and dogs), or in cold weather.

Closely related to hair are nails, claws, hoofs, and horn.*

* Here is intended such horns as those of the cow, sheep, antelope, and rhinoceros ; the horns of the deer are true bone.

In fact these structures must be regarded as hairs united throughout their length.   At other times a similar consolidation of hair gives rise to protective scales covering the body, as in the case of the pangolins.

The bodies of the vertebræ usually have flat faces, and the vertebral column in most forms can be divided into five regions—cervical, thoracic, lumbar, sacral, and caudal. The cervical vertebræ occur in the neck; they bear no ribs, and, except in three rare forms, they are constantly seven in number, the long-necked giraffe and the short-necked whale having the same number of these bones.   The thoracic vertebræ are more variable in number.   They bear ribs, some of which extend downward and unite with the breastbone or sternum.   Between the thoracic and pelvic regions occur the ribless lumbar vertebræ, while the sacral vertebræ are those which unite with the pelvic bones.   The caudal vertebræ are found in the tail.   In the whales only cervical and thoracic vertebræ can be distinguished, since the absence of a pelvis in these forms allows no line to be drawn between lumbar, sacral, and caudal regions.

In the skull there is a tendency for bones which are distinct in the fishes and reptiles to fuse with each other, so that the number of distinct elements is considerably reduced.   The skull is borne on the first cervical vertebra, upon which it slides by means of two rounded surfaces or **condyles.**   The lower jaw articulates directly with the skull, and is never suspended by a quadrate bone, as in the forms already studied.

The fore limbs are always present; the hind limbs are absent in the whales and manatees, being represented in a few forms by one or two bones imbedded in the muscles of the trunk.   Except in the Monotremes (p. 102), the coracoid does not occur as a distinct bone, but as a small prominence joined to the shoulder-blade (**scapula**), while in many the collar-bone (**clavicle**) also is lacking. The feet have typically five toes, but not infrequently this number is reduced by a disappearance of the outer digits, the reduction reaching its extreme in the cow, which has but two, and the horse, which walks upon the tip of its middle toe.

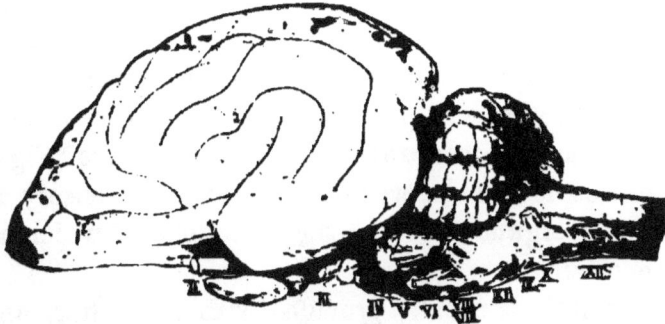

FIG. 37.—Brain of Dog. (After Weidersheim.) II–XII, the cranial nerves (see page 94).

The most marked characteristic of the nervous system is the great relative increase in size of the cerebrum, and, to a less extent, of the cerebellum; the optic lobes and the medulla, so prominent in the lower forms, being overshadowed by these parts. The cerebrum is the seat of intelligence, and this increase in size is correlated with the higher mental powers of the mammals. Microscopic study of the brain has shown that this organ is composed of two different portions, called, according to their colors, white and gray, and that the gray matter is the true brain substance, while the white is composed of nerve-cords to transmit nerve impulses. The gray matter is on the outside of the cerebrum, hence the larger the brain the more surface it has, and consequently the more gray matter it can have. In the higher mammals the amount of surface of the cerebrum is greatly increased by folds or convolutions, and the extent and complexity of these convolutions correspond well with the intelligence of the form.

In the eyes the nictitating membrane or "third eyelid" of the bird is reduced to a small fold at the inner angle of the eye. Except in the whales, and some seals, moles, etc., external ears are developed, while the internal parts of the ear become considerably modified. Thus the quadrate and one other bone pass in to the middle ear, where they, together with a third bone, form a chain to convey sound-waves to the sensory portions. In the inner or sensory portion a spiral outgrowth, the cochlea, occurs (Fig. 64), and in this is a most wonderfully intricate sensory apparatus—**the organ of Corti**—the functions of which are as yet uncertain.

The mouth is usually provided with fleshy lips, and all mammals, except some edentates and whales, have teeth. These teeth are always confined to the jaws (*cf*. Fishes, p. 10), being inserted by one or more roots into sockets in the bone. Some mammals have but a single set of teeth throughout life, but the majority have a first or **milk dentition**, which is soon lost and replaced by a **permanent dentition**. Occasionally, as in the sperm-whale, etc., all the teeth are similar in shape, but usually several different kinds occur, the extreme being reached when four types are present—incisors, canines, premolars, and molars. The **incisors** have but a single root, and are found in the premaxillary bone and in the corresponding position in the lower jaw. The first teeth in the maxillary, if single-rooted and pointed, are called **canines**; and behind these come the molars, with two or more roots. These in turn are subdivided into **premolars**, which appear in both milk and permanent dentitions, and molars proper, which occur only in the permanent set. The number of teeth and their arrangement vary considerably in different mammals, and the characters which they furnish are of great value in grouping the various forms. To express these characteristics briefly a **dental formula** has been introduced, in which the different kinds of teeth are indicated by initials, while the number in either jaw is represented by a figure above or below a horizontal line. Thus the permanent dentition of man is expressed thus: $i\ _2$, $c\ \frac{1}{1}$, $pm\ \frac{2}{2}$, $m\ \frac{3}{3}$; which indicates that in man there are two incisors, one canine, two premolars, and three molars in each half of each jaw. The pig has, $i\ _3$, $c\ \frac{1}{1}$, $pm\ \frac{4}{4}$, $m\ \frac{3}{3}$; the cow, $i\ \frac{0}{4}$, $c\ \frac{0}{1}$, $pm\ \frac{3}{3}$, $m\ \frac{3}{3}$, incisors and canines being absent from the upper jaw.

The body-cavity is divided by a transverse muscular partition, the **diaphragm**, into two chambers—an anterior **pleural** cavity containing the heart and lungs, and a posterior **peritoneal** cavity in which is situated the stomach, liver, intestine, etc.

The heart, placed a little to the left of the median line, is four-chambered, having, like that of the birds, two auricles and two ventricles. Of these the auricle and ventricle of the right side receive the blood from the body and send it to the lungs, while those of the left side take the blood as it comes from the lungs and send it through the aorta to all parts of the body. The aorta, which bends backward and to the left, represents the left arch of the fourth pair of the primitive branchial vessels, the right of the same pair being partially represented in the artery (subclavian), which carries the blood to the right fore limb —a condition just the reverse of what occurs in the birds. The fifth pair of arches form part of the arteries (**pulmonaries**) which convey blood from the heart to the lungs. The blood of the mammals differs from that of all other forms in that the red corpuscles are circular in outline and are not nucleated.

The monotremes form the only exceptions to the statement that the mammals bring forth living young. These lay eggs, one species having the eggs about the size of a pigeon; but the young which are hatched from these eggs are nourished by milk secreted by the mother, as is the case with all other mammals.

The Mammalia are divisible into three large groups or subclasses: Monotremata, Marsupialia, and Placentalia.

SUBCLASS I.—MONOTREMATA.

This subclass contains three or four species of animals which are found only in Australia and its immediate neighborhood. They present resemblances to the birds, or

FIG. 38.—Duckbill (*Ornithorhynchus paradoxus*). From Lütken.

better, to the reptiles in the following points, in all of which they differ from the other mammals: They lay eggs; they have well-developed coracoid bones; and reproductive and excretory organs empty into the posterior portion (**cloaca**) of the intestine, and thence pass by a common opening to the exterior.

The monotremes include the duckbill and the spiny ant-eaters. The duckbill is an aquatic animal, and receives its common name from the fact that it has a horny bill much like that of the duck. It lives in burrows in the banks of streams, and feeds on beetles, shrimps, etc., which it catches in the water and crushes with its horny teeth, its true teeth being lost at an early age. The spiny ant-eaters resemble the duckbill in their burrowing habits, but they live exclusively on the land, where they feed on ants. They are, like the true ant-eaters (p. 106), entirely toothless, and receive the adjective spiny of their common name from the fact that their hair takes the shape of long stout spines, recalling those of the porcupines.

SUBCLASS II.—MARSUPIALIA.

This subclass receives its name from the fact that in the female a curious pouch or marsupium is developed on the

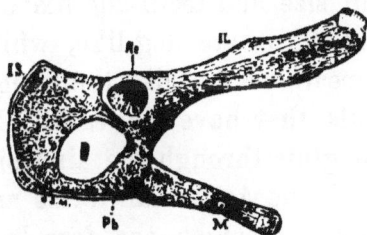

FIG. 39.—Pelvis of Opossum. (After Minot.) *M,* marsupial bone; *il,* ilium; *is,* ischium; *p,* pubis.

FIG. 40.—Opossum (*Didelphys virginiana*). After Audubon and Bachman.

lower surface of the body of the female, in which the young are placed by the mother immediately after birth, and where they remain until able to take care of themselves. This pouch is supported by a pair of bones which extend forward from the pelvis—the marsupial bones,—and these, as well as a peculiar inbending of the angle of the lower jaw, serve at once to distinguish any marsupial skeleton. Were these the only characters to be considered we should not be warranted in placing these forms in a subclass by themselves, but there are other characters connected with reproduction which justify this course. The living marsupials have a peculiar distribution: they are restricted to warmer America and the chain of islands extending from Australia to Celebes. Fossil forms are found in Europe as well.

The American species are all opossums—forms with prehensile tails — and have given rise to the expression "playing 'possum," from their habit of feigning death when disturbed. Their food is chiefly insects, but birds, eggs, etc., are not despised.

Australia is the real home of the marsupials; indeed at the time of its discovery this continental island contained only marsupials, if we except the dingo, or native dog. In this region are found forms which recall animals of different groups occurring in other parts of the world. Thus the wombat resembles in size and teeth the beaver; the thylacines in habits and in form are dog-like, while the phalangers in size and appearance are like the flying squirrels, and, like those animals, they have that same fold of skin which enables them to glide through the air from tree to tree. Most familiar of all the Australian forms are the large grass-eating kangaroos, in which the fore legs have become almost useless for locomotion, the animal jumping with its hind legs, and when resting, supporting itself upon these members and its enormously developed tail. There are also fossil marsupials in Australia, some of them of enormous size. Thus Thylacoleo was as large as a lion, while Diprotodon had a skull three feet in length and a thigh-bone two feet from tip to tip.

SUBCLASS III.—PLACENTALIA.

The great majority of mammals belong in this division. They are marked off from the other subclasses by the absence of those characters which have been mentioned as distinguishing these, as well as by a structure now to be mentioned. These mammals are not born until their internal organization has been well advanced; and in order that they may be supplied with nourishment a peculiar vascular structure is formed,—the **placenta**,—by means of which blood is brought to the growing embryo. Such a structure is lacking in both monotremes and marsupials. The Placentalia are divided by details of structure into many groups or orders, eleven of which are represented in the world to-day.

ORDER I.—EDENTATA.

The edentates, the lowest of the placental mammals, receive their name from the fact that incisor teeth are

FIG. 41.—Nine-banded Armadillo (*Dasypus noremcinctus*). From Lütken.

always lacking, while in the ant-eaters no teeth occur. The feet are armed with strong claws. The group is a tropical one, and has its greatest representation in America. Here belong the armadillos, in which the deeper layer of the skin becomes converted into bone, forming an armor over the body. In the fossil Glyptodon this armor formed one solid piece, enclosing the trunk much like the armor of a turtle; but in the living forms it becomes broken into several transverse bands, which move upon each other, so that the animal can coil itself into a ball.

The sloths are larger forms which, back downward, crawl with the slowest motions along the branches of the trees, holding themselves by their hook-like claws. Upon the ground they walk with difficulty, their long claws being in the way. In geological times there were forms

FIG. 42.—Pangolin (*Manis longicaudata*). From Monteiro.

allied to the sloths, but of much larger size. One, the Megatherium of South America, had a skeleton 18 feet in length. Another form found in North America receives interest from the fact that it was first described by Thomas Jefferson.

The ant-eaters are true edentates in that they are wholly without teeth. As their name implies, ants form the chief part of their food; their claws are well adapted for digging

into the nests, the tongue is very long and extensible, while
the salivary glands pour out a thick, sticky secretion which
fastens the ants to the tongue. The true ant-eaters are
natives of South America, but in Africa and India are
allied forms with teeth, which also feed upon ants.
Among these are the pangolins, in which the whole upper
surface of the body is covered with scales, arranged somewhat like those of a pine-cone. These scales, as already
mentioned (p. 98), are to be regarded as modified hair.

ORDER II.—RODENTIA (The Gnawers).

The rodents are the gnawers, the well-known abilities of
rats and mice in this direction being shared by all members of the order. They have no canine teeth; the molars

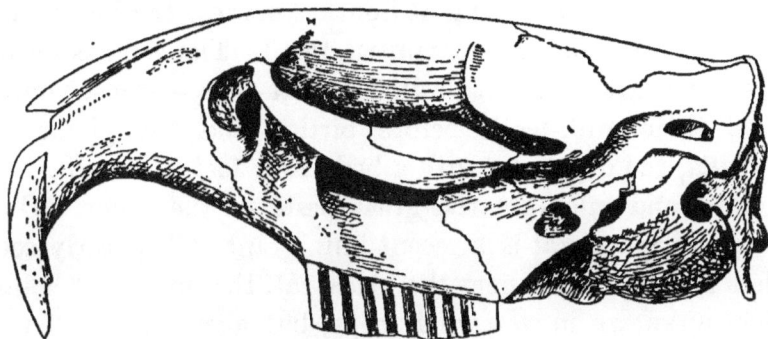

FIG. 43.—Skull of muskrat (enlarged), showing the gnawing incisors and absence
of canines.

are usually $\frac{3}{3}$, while the incisors vary between $\frac{2}{2}$, $\frac{4}{2}$, and $\frac{6}{2}$.
These incisors demand a moment's attention. These teeth
have **persistent pulps**, i.e., they continue to grow throughout life. As fast as they wear away they are renewed from
below. In each incisor two parts can be distinguished:
the anterior face of the tooth is covered with a very hard
layer (**enamel**), while the posterior surface is composed of
a much softer **dentine**. This dentine wears away much

faster than the enamel, and the result is that the teeth are constantly kept at a chisel-edge.

Lowest of the rodents come those forms familiarly known as hares and rabbits, with disproportional hind legs and long ears. The distinction between the two—hares and rabbits—is very slight, the true rabbit being a native of southern Europe. All the rest are hares. In America, however, the term rabbit is usually restricted to the small burrowing forms.

The porcupines, with some of their hair changed to long sharp spines,—efficient weapons of defence,—come next. These occur in both hemispheres, but the American forms are mostly arboreal, while those of the Old World burrow. Allied to them in structure, but differing in fur, are the chinchilla and the coypu of South America, the latter furnishing the well-known "nutria fur." The same country furnishes the stupid, so-called guinea-pigs,—whose young shed their milk-teeth before birth,—and the giants of rodents, the capybara, with a body four feet in length.

Rats and mice are the great pests of the order. Our common brown rat is a recent immigrant. The early settlers brought with them the black rat, the brown rat being then unknown in western Europe, but about 1720–30 the latter came west from the Volga region, and gradually spread all over western Europe and then over America, the black rat disappearing before the invader. There are many rat-like forms, among them the lemmings of the Arctic regions, vast hordes of which occasionally overrun Norway; the dormice, which hibernate in winter; the gopher and pocket-rats, which burrow through the soil in the Western States; the familiar muskrat, and the less familiar jumping mice, which resemble the kangaroos in their locomotion.

Another series of rodents contains the beaver, common to the Old World and the New, which furnishes furs of great value. These live most of their lives in the water, building dams so that they may always have plenty of it; while their near relatives, the woodchucks, and their western representatives, the prairie-dogs, have no such dependence upon water. Highest of all the rodents are the ground - squirrels, the true squirrels, and the flying squirrels.

ORDER III.—INSECTIVORA (Insect-eaters).

These are small forms, in which all four types of teeth are developed, and which are marked off from all other orders by characters rather difficult of expression. As their name implies, they feed largely upon insects, but worms and other small animals are not despised. The species are largely tropical, but the shrews and moles are found in cooler climates. Most of the species are nocturnal and burrowing animals, consequently their eyes are small, while their fore legs are adapted for digging.

ORDER IV.—CHEIROPTERA (Bats).

The bats are the only mammals which truly fly. In the case of the flying squirrel and the rest, the animals glide through the air on the plane formed by the lower surface of the body, the tail, and the broad membrane which extends between the limbs; and they can never ascend to the level from which the flight started. With the bats, on the other hand, there are no such limitations to the flight. The wing in the bats consists of a very thin membrane supported upon a framework composed of the body and the bones of the fore limbs. These latter are elongated, four of the fingers excessively so ; and between these fingers and extending back to the body and the hind limbs is the web of the wing. The thumb, however, is not involved in the wing, but forms a claw of great use in supporting the body, although when at rest they usually hang by the five claws of the hind feet.   The jaws are provided with incisors, canines, premolars, and molars.   Bats are social animals, occurring in large numbers in caves, deserted buildings, and

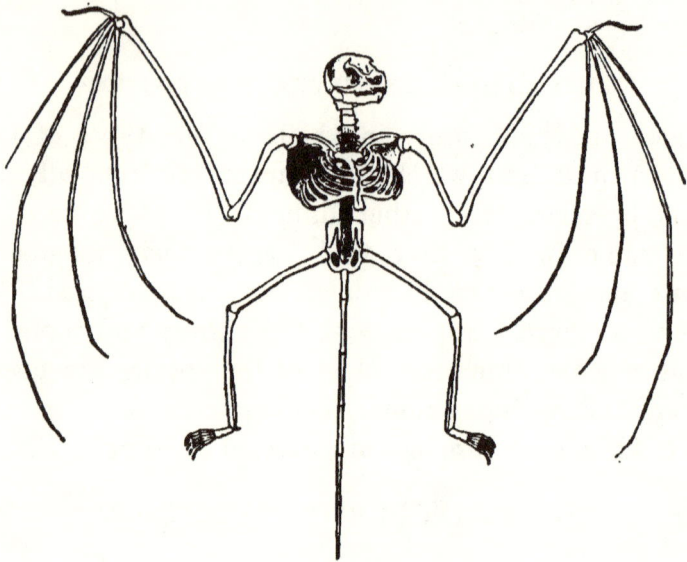

FIG. 41.—Skeleton of bat.

the like, where they spend the day, and it is remarkable that these colonies are usually entirely male or female.   In a rough way the bats may be divided into fruit-eating and insect-eating forms, their habits being correlated with peculiarities of structure.   To the fruit-eating species belong the large bats of the East Indies known as flying foxes.   All of our bats are insect-eating.   Some of the South American bats (not the one called the vampyre by Linné) are known to suck the blood of other mammals.

In the four orders Edentates, Rodents, Insectivores, and Bats the surface of the cerebrum is smooth; in all the remaining orders it is at least fissured, and in most it is convoluted (see Fig. 37), this increase in surface reaching its greatest development in man. Since this line of division corresponds in a way with the intelligence of the forms, the four orders already mentioned are grouped together as Ineducabilia; the others are associated as Educabilia.

ORDER V.—CETE (Whales).

The whales have a fish-like body, the resemblance being frequently heightened by the development of a dorsal fin ; and yet in all points of structure they are mammals. The anterior limbs contain the same bones (except that the number of joints in the fingers may be increased) as do our

FIG. 45.—Pigmy whale (*Kogia floweri*). From Gill.

own, but the whole has been modified into a "flipper" for use in swimming. The hind limbs are absent externally, but imbedded in the flesh on either side is a bone, variously interpreted as a part of the pelvis or as the bone of the thigh. The body terminates in a bilobed caudal fin ("flukes"), but this, instead of being vertical, as in the fish, is horizontal. All of the whales have teeth in the young stages ; some retain them through life, while others lose them long before maturity, sometimes even before birth. The stomach is remarkable for having several (4–7) chambers, this complication recalling the condition in the cow.

According to the presence or absence of teeth the living whales are divided into two groups. In some of the toothed whales but two teeth are present; others may have a large number ; and usually these cannot be well distributed among incisors, canines, etc., as all are essentially alike in size and shape. In the male narwal, however, one of the upper teeth on one side (apparently a canine) grows straight forward into a long twisted spear eight or nine feet in length, while the other teeth disappear at an early age. The killer-whales are comparatively small, but are among the most voracious of mammals, not hesitating to attack the largest whales. Here also belong the blackfish, porpoises, and dolphins. The sperm-whales are larger, and have no teeth in the upper jaw, while the lower jaw is abundantly supplied. They derive their common name from the spermaceti which they produce. This is a solid granular substance found in the "case," a cavity occurring on the right side of the front of the head between the skin and the skull.

The toothless whales are also known as whalebone whales, from the fact that they bear upon the lower sides of the upper jaw hundreds of long parallel plates of so-called whalebone or baleen. These plates are fringed at the end, and the whole apparatus forms an efficient strainer, used in separating the small animals upon which these whales feed from the surrounding water. It is among these whalebone whales that the giants among mammals occur.

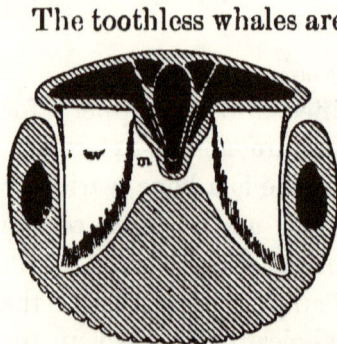

FIG. 46 —Section through the head of a whalebone whale (after Boas), showing how the plates of baleen (*w*) are arranged on either side of the mouth-cavity (*m*) The true bones are shown black.

The right whales of the Arctic

seas reach a length of sixty feet, the razor-back whales are still larger, while the sulphur bottoms and silver bottoms (so called on account of the color of the lower surface) attain a length of from 90 to 95 feet.

ORDER VI.—SIRENIA (Sea-cows).

These are whale-like animals, with the same flippers and the same horizontal tail, but they differ from the whales in the possession of an evident neck, and of sparse hair or bristles all over the body. Besides these features all, except the extinct Rytina, have flat-crowned molar teeth. The living forms are very few. Rytina, which lived near Bering Strait, was exterminated in the last century. The dugong is the representative of these forms in the Indian Ocean, while the three species of manatees come, one from Africa, the other two from the eastern coasts of America. All the sea-cows are vegetable feeders, living upon sea-weed or, in the case of the manatees, upon the plants found in fresh-water streams as well.

ORDER VII.—PROBOSCIDIA (Elephants).

The elephants are the giants among the land mammals. They have five toes, each encased in its own hoof; they have no incisors in the lower jaw, while the pair in the upper jaw are developed into large tusks. Canines are lacking, but there are seven molars in each half of each jaw. These molars are flat-crowned, the surface of the crown being crossed by several ridges of harder enamel. Only two, or at most three, of these molars are in use at once, but as the old ones wear out they drop out at the front of the jaw, and are replaced by new ones from behind until the seven are gone. The skull is enormous, but it is comparatively light on account of the numerous cavities in the bone.   Most striking of all is the proboscis, which is

FIG. 47.—A manatee (*Trichechus americanus*) feeding.   After Elliott.

merely an enormously developed nose, with capacities which only one who has studied an elephant can realize.   The skin is almost entirely naked, hairs being scarce, and on the tail taking the shape of long wiry bristles.

To-day two species of elephants exist, one having its home in India, the other in Africa. In the later geological ages there were several others, one having lived in America and others in Europe. Towards the end of the last century remains of hairy elephants—even the flesh being preserved —were found imbedded in the ice in northern Siberia Allied to the elephants were the somewhat larger mastodons, in which the molar teeth bore conical cusps, while the tusks were frequently enormous. Some mastodons had incisors in the lower jaw as well.

ORDER VIII.—HYRACOIDEA (Coneys).

This order contains but two or three species, distributed from Syria south into Africa. In having long curved incisors and absence of canines they recall the rodents, in other points their structure is like that of the rhinoceros, while the foot-pads on their feet recall those of the cat or dog. The hyrax of Syria is probably the coney of the Old Testament.

ORDER IX.—UNGULATA (Hoofed Animals).

To this order belong the great majority of important mammals. They are herbivorous, usually of large size, and lack collar-bones. The feet are used solely in walking, and not in prehension, each toe having its tip enclosed in a horny hoof, and in living forms there are never more than four toes developed on a foot. The living ungulates are arranged in two series, according to the number—even or odd—of toes upon their hind feet. The odd-toed forms are called PERISSODACTYLA, the even-toed are ARTIODACTYLA.

To the perissodactyls belong, of living forms, the tapirs, rhinoceroses, and horses. The tapirs live in the forest regions of the tropics of both continents. They have a hog-like body, large prehensile upper lip; teeth, $i\frac{3}{3}$, $c\frac{1}{1}$, pm $\frac{4}{4}$, $m\frac{3}{3}$; while their fore feet have four toes, the hind feet three. Yet, although the fore feet have an even number of toes, these are not symmetrically arranged, as in artiodactyl forms, the pig for example, but one (third) is enlarged and bears most of the weight of the body.

The rhinoceroses have three toes on each foot; the skin

Fig. 48.—Sumatran rhinoceros (*Ceratorhinus sumatrensis*).   From Lütken.

is extremely thick; the snout bears one or two well-developed horns, in which there is no bony core; and canine teeth are not developed even in the young. There are six species known, those occurring in Africa having two horns, while in the East Indies are both one and two-horned forms.

In the horses the reduction of toes has gone still farther,

there being but one (the middle or third) in each foot. In
the skeleton, however, traces of two more can
be found in the "splint-bones," two small
bones occurring alongside the large "cannon-
bone." All of the existing horse-like forms
have the teeth i $\frac{3}{3}$, c $\frac{1}{1}$, p $\frac{4}{4}$, m $\frac{3}{3}$, and all are
natives of the Old World, none existing in
America at the time of its discovery. All
evidence goes to show that the home of the
domestic horse was in Central Asia, and indeed
four different species of horse run wild there
to-day. The asses have their centre around
the eastern end of the Mediterranean, while
the zebras or striped horses are all African. In
geological time, however, America had horses,
and the fossils in our Western States give the
history of the race from small forms about the
size of a fox, and with three toes behind and
four in front; later, those as large as a sheep,
with three functional toes in each foot; and

Fig. 49.—Foot of horse, showing the splint-bones (second and fourth toes) at 8; 3, third toe.

still later, three-toed forms as large as a donkey. In
domestication horses vary extremely in size as in other
respects.

Lowest of the artiodactyls, or even-toed ungulates, come
the two species of hippopotamus, in which there are four
toes, large canine teeth, and a huge, clumsy body, sometimes fourteen feet in length. In the pigs the canines are
still large, and the toes are four in number, but the outer
ones are lifted above the ground so that they are useless as
organs of locomotion. Our domestic swine have descended
from the wild-boars of Europe. In the warmer parts of
America the peccaries represent the group.

The hippopotamus and the pigs have the axis of the

foot passing up between the middle toes; in other words, they have cloven hoofs.   In all other artiodactyls the cloven hoof occurs, and besides, they chew the cud, and hence they are associated as a group of ruminants.   The stomach is divided into four chambers, and when a cow, for instance, feeds, it swallows the grass without chewing it.   It passes down to the first stomach and thence to the second.   In these it becomes mixed with digestive fluids and softened.   It is then brought up into the mouth, thoroughly chewed, and again swallowed.   This time it passes into the third stomach, and from this into the fourth, and so into the intestine.

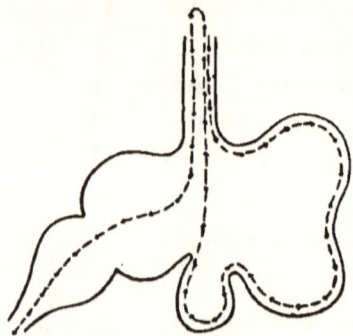

Fig. 50.—Diagram of the stomach of a ruminant.   The dotted line shows the course of the food.

To the ruminants belong the most valuable domesticated animals.   In South America are found the llamas and alpacas, which were the cattle and beasts of burden of the ancient Peruvians; while in Asia and Africa the camels, in part, take their place.   Two kinds of camels occur, one with one and the other with two humps upon the back. These humps are merely large masses of fat.   Some forty years ago the United States Government introduced some camels into our southwestern territory, and the descendants of these are still to be found in Arizona.

We associate together under the common name of deer all those ruminants which have horns consisting of solid bone.   These horns are annually shed and grow out anew each year, usually increasing in size with the age of the animal.   When first formed the horns are covered with a thin skin with short hairs.   The horns in this condition are said to be in the velvet. When the horn is fully formed the skin dies and is worn off. In some deer horns are borne only by the male, but sometimes both sexes, as with the reindeer, are provided with them. The long-necked giraffes are closely related to the deer.

In other ruminants the horns are never shed. In these the horns consist of a central core of bone, covered on the

FIG. 51—Prong-horned antelope (*Antilocapra americana*).

outside with a horny structure—in reality modified hair (p. 97). Here belong our domestic cattle, which are believed to have arisen from four different kinds, which formerly were wild in Europe. This wild stock is almost extinct. One of these forms at least was closely similar to our American bison, which has so nearly approached extinction from the desire for "buffalo" robes. The true buffalo are all natives of the Old World, and occupy a position between the ancestors of domestic cattle and the long series of forms grouped together as antelope, most of which belong to Africa, but which are represented in America by the prong-horned antelope of our Western States. Other members of the same group with permanent horns are the sheep and the goats, the series ending with the so-called musk-ox of the arctic regions, a form nearer the goats than to the domestic cattle in its structure.

As a whole, we may say that in points in structure— especially in the characters of feet and teeth—the group of ungulates are among the most specialized of the mammalia, the whales, bats, seals, and possibly the elephants alone excelling them in this respect.

Order X.—Carnivora (Beasts of Prey).

The beasts of prey are specialized in the direction of flesh-eating. Their bones are slender, but strong; their feet (usually five-toed) are furnished with claws; while on the top of the skull is a crest for the attachment of the strong muscles of the jaws. All four kinds of teeth are present, and one of the molars or premolars is flattened vertically, so that, meeting its fellow of the opposite jaw, it cuts like a pair of shears. In the lower mammals we find the lower jaw so hinged upon the skull that it can move back and forth in grinding the food. In the carnivores, on the other hand, no such motion is possible.

The carnivores are divided into two groups, one embracing the typical land-inhabiting forms; the other, which includes the walrus and the seals, is modified for an aquatic life; the differences being most marked in the structure of the appendages. In the first group the legs are elongate and the toes are distinct, whence the name Fissipedia; while in the other division (Pinnipedia) the legs are shortened, the fingers are webbed, and the feet are thus effective paddles.

Lowest of the fissipedia are the bears and their allies, in which the whole sole of the foot is applied to the ground in walking, and hence are called **plantigrade**, in opposition to those **digitigrade** forms, like the cat and dog, which walk upon the tips of their toes. The bears are widely distributed over the earth, America having at least three species. The racoon is distributed throughout the United States, and in tropical America is represented by that exceedingly interesting animal, the coati.

Another group of carnivores includes the otters, mink, ermine, sable, and marten—all of which are valuable for the furs which they afford,—as well as the weasels and ferrets, and the well-known skunks. These are partly plantigrade, partly digitigrade.

The dogs, foxes, wolves, and jackals are all digitigrade. They have the teeth, $i\ \frac{3}{3}$, $c\ \frac{1}{1}$, $pm\ \frac{4}{4}$, $m\ \frac{2}{3}$. Foxes and wolves are wild, and many believe that our domestic dogs have descended from some wolf stock; but others think that dogs and wolves are distinct, and even that our common dogs represent several originally distinct kinds or species.

The hyænas are intermediate between the cats and dogs in many respects. They have the back teeth fitted for crushing. In the cats, of which there are more than fifty species, the teeth are usually $i\ \frac{3}{3}$, $c\ \frac{1}{1}$, $pm\ \frac{3}{2}$, $m\ \frac{1}{1}$, while the claws are retractile into sheaths. Our domestic cat apparently had its origin in Egypt, while ancient Greece and Rome lacked our familiar puss, its place being taken by domesticated martens. Among the cats the tiger, lion, panther, leopard, and puma rank first, and with them are associated the wildcats and lynxes.

In external form the seals and walruses have little resemblance to the other carnivores, but in structure, and especially in their skulls, there is great resemblance—to the

bears and otters in particular.   As has been said, their feet
are modified into paddles, and only the distal region is distinct from the body.   Lowest are the large walruses, of
which there are two species in northern seas, in which the
upper canines are enormously developed.   They can use
their hind feet in walking.   The eared seals are so-called
because they have small external ears.   The largest of these
are the sea-lions, but the most valuable are the fur seals, of

Fig. 52.—The harbor seal (*Phoca vitulina*).   After Elliott.

which two species are known.   The one which occurs in the
southern hemisphere has been almost exterminated, while
the Alaskan species is rapidly following the same road.

The true seals lack all external ears, and since their skins
are less valuable, a longer lease of life seems assured them.
They occur on all shores, and from their fish-eating habits
are frequently a nuisance to fishermen.

Order XI.—Primates.

The term Primates is given to that group which includes the monkeys, apes, and man, from the fact that they are the first or highest group in the animal kingdom. Collar-bones are always present; the feet are very primitive, and the fingers and toes are armed with nails, claws but rarely occurring. Intelligence, not structure, assigns them the leading place.

Lowest come the group of lemurs or "half apes," which have their metropolis in Madagascar, but have relatives in Africa and in the East Indies. They are largely nocturnal, and eat fruit or insects or other small animals. They are noticeable from the fact that the second finger is provided with a claw.

The marmosets are small squirrel-like forms found in South America. They are provided with claws on all digits except the great toe, and the tail is incapable of grasping, while the thumb is scarcely capable of being opposed to the fingers.

The remaining American monkeys—the howlers, sapajous, spider-monkeys, and the like—have a broad septum of the nose, causing the nostrils to be wide apart; the thumb is scarcely opposable, and in some is lacking; while the teeth differ from those of the Old World monkeys, and of man, in having pm $\frac{3}{3}$. Many have a prehensile tail.

The Old World monkeys have the nostrils closer together, the thumb as well as the great toe is opposable, and the tail never takes the place of a fifth hand. In their teeth they resemble man: i $\frac{2}{2}$, c $\frac{1}{1}$, pm $\frac{2}{2}$, m $\frac{3}{3}$. The baboons, distributed across Asia and Africa, have large cheek pouches for the storage of food, etc., and naked callous patches on which they sit.   Some have long tails, others no tails at all.   The macaques and mangabeys are allied Asiatic forms.

In the anthropoid apes tail, cheek pouches, and callous spots are lacking; as the name indicates, they are man-like.   There are three of these.   The orang-utan (the name is Malay for Man of the Woods) lives in Borneo and Sumatra.   The chimpanzee and the gorilla are African.

FIG. 53—Chimpanzee (*Troglodytes niger*).   After Brehm.

Each of these has certain points in which it is more like man than are the others.

The highest mammal is man, who differs from the other primates less in structure than in intelligence.

COMPARISONS.

With five columns, one for fish and dogfish, one for frog, one for turtle and snake, and one each for bird and rat, answer the following questions:

(1) Is the body bilaterally symmetrical ?

(2) Are paired appendages present ?

(3) How many nostrils ?

(4) How many eyes ?

(5) How many ears ?

(6) Are the skeletal parts external or internal ?

(7) Is the vent dorsal or ventral ?

(8) Is there a skull ?

(9) In what plane do the jaws move ?

(10) Is the back-bone a single structure ? If not, of what is it composed ?

(11) Are both shoulder and pelvic girdles present ?

(12) On what side of the alimentary canal is the central nervous system ?

(13) What parts are found in the central nervous system ?

(14) To what organs do the first pair of nerves go ?

(15) To what organs do the second pair of nerves go ?

(16) How many and what parts do you find in the brain ?

(17) Are there cavities inside the brain ?

(18) Is there a peritoneal cavity ?

(19) In what way is the alimentary canal supported ?

(20) Do you find in each form liver, spleen, and pancreas ?

(21) In what part of the cavity do you find the kidneys ?

(22) What cavity surrounds the heart ?

(23) What chambers do you find in the heart ?

(24) Is the heart dorsal or ventral to alimentary canal ?

(25) Is the aorta dorsal or ventral to alimentary canal ?

(26) What vessels carry blood to the head ?

(27) Are the respiratory organs connected, either directly or indirectly with the alimentary canal ?

(28) Are any parts (if so, what) repeated one after another in the body ?

(29) Draw a diagram of a transverse section through the body in the region of the heart, showing the heart, spinal cord, œsophagus, vertebra, aorta, and body-walls.

(30) Draw a similar section through the kidneys, showing the peritoneal cavity, intestine, mesentery, spinal cord, kidneys, aorta, vertebra, etc., and the body-walls.

All of the forms so far studied or described are associated together as a group or branch—Vertebrata—the name of which implies that they all possess a " back-bone " composed of separate bones or vertebræ. This one character of itself would hardly warrant this grouping, especially since some forms have the vertebræ but feebly developed, while in other features they are closely similar to those with a well-developed back-bone. This presence of vertebræ is closely associated (correlated) with other features of equal or even of more importance, and it is this totality of similarity that justifies the group.

All vertebrates have an inner supporting skeleton, and a few forms, like the turtles, have in addition an external skeleton derived from the skin. The internal skeleton, for convenience of treatment, may be divided into one portion lying in the axis of the body, and a second portion pertaining to the limbs and appendages. Besides these there is a third part, the **visceral skeleton**, developed in connection with the gills.

The axial skeleton consists of the vertebral column (backbone), the skull, and the ribs. In all vertebrates, at least in the young stages, a solid rod of cartilage runs through the body between the central nervous system and the alimentary canal. In front it terminates near the middle of the brain; behind it runs to the end of the body. This rod is the **notochord**. In the higher vertebrates it disappears long before the animal becomes adult: but in the

127

lower, as in the sharks, it can be recognized throughout
life. This notochord is enveloped in a membranous **notochordal sheath,** and in this sheath are formed rings of
cartilage which give rise to the bodies (**centra**) of the vertebræ. Between these rings no cartilage is formed and hence
the whole column is jointed and flexible. In the sharks
these rings and other parts of the skeleton remain cartilaginous; in other vertebrates any or all may be converted into bone. In a typical vertebra, for instance, in
the tail of a fish (p. 14), outgrowths from the centrum
occur above and below, forming two arches. The upper of

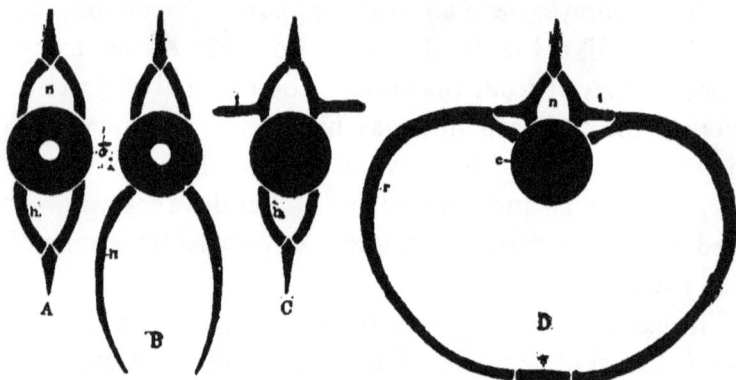

FIG. 54.—Different vertebræ and connected structures. *A*, in tail region
of teleost; *B*, in body region of teleost; *C*, in tail region of salamander;
D, in mammal; *c*, centrum; *h*, hæmal arch (rib in *B*); *n*, neural arch;
r, rib; *s*, sternum; *t*, transverse process.

these (**neural arch**) encloses the spinal cord, the lower
(**hæmal arch**) extends around the blood-vessels of the tail.
Farther forward, in the trunk region of the bony fish, the
two halves of the hæmal arch do not meet below, but form
slender threads (**ribs**) which support the flesh around the
viscera. In the forms above the fishes an outgrowth
(**transverse process**) may rise on either side of the vertebral
centrum, and the ribs, when they occur, are continuations of
these transverse processes, and have nothing to do with the

hæmal arches. Hence it follows that the ribs in a fish and those in a higher vertebrate—a bird or man, for example— are not identical; i.e., are not homologous. The centra of the vertebræ may be hollow at either end (**amphicœlous**) as in fishes, or they may be hollow behind and rounded in front (**opisthocœlous**) as in the salamanders; or again they may be rounded in front and concave behind (**procœlous**) as in many reptiles; or lastly, they may have flat surfaces, as in most mammals.

The vertebral column is capable of division into regions.

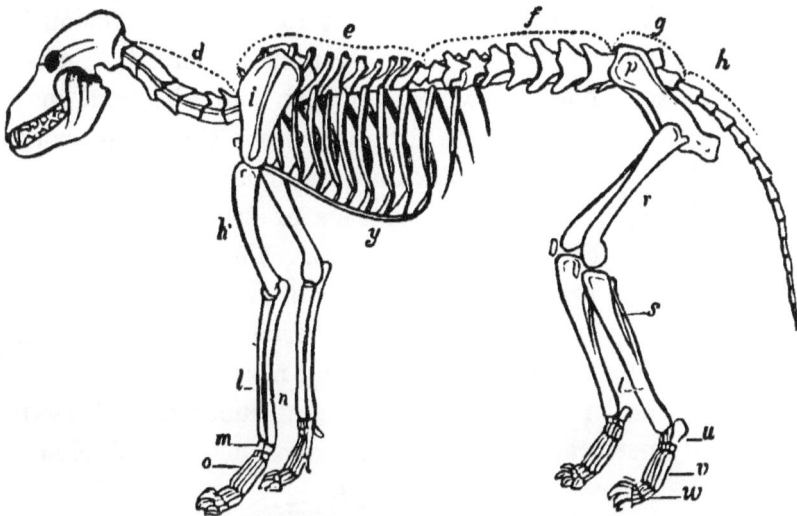

FIG. 55.—Diagram of the skeleton of a mammal, showing regions of vertebral column, etc. *d*, cervical; *c*, thoracic; *f*, lumbar; *g*, sacral; *h*, caudal vertebræ; *i*, scapula; *k*, humerus; *l*, radius; *m*, carpus; *n*, ulna; *o*, metacarpus; *p*, pelvis; *r*, femur; *s*, fibula; *t*, tibia; *u*, tarsus; *v*, metatarsus; *w*, phalanges; *y*, sternum.

In the fishes there are two of these, trunk and caudal, the former being distinguished by bearing ribs. In the Batrachia a cervical region is distinguished from the trunk by the absence of transverse processes from its single vertebra, while the caudal is separated from the trunk by a sacral region, the vertebra of which is connected with the bones (girdle) supporting the hind limbs. In the higher vertebrates the trunk vertebræ can be divided into thoracic and lumbar regions, the former with, the latter without, ribs.

As we have just seen, there may be two kinds of ribs—those of fishes and those of the higher vertebrates. In reptiles, birds, and mammals the ribs of one side fuse at their ventral ends with their fellows of the opposite side. The fused regions separate from the ribs and unite together, giving rise to the breast-bone or sternum. In some sterna the separate elements can be traced; in others the fusion is complete. The sternum in the Batrachia has no connection with the ribs, and may therefore be different from the breast-bone in Sauropsida and Mammalia.

The skull consists of two portions: the cranium and the face. The former affords protection to the brain and support to the organs of sense; the facial portions cluster around the mouth and nose.

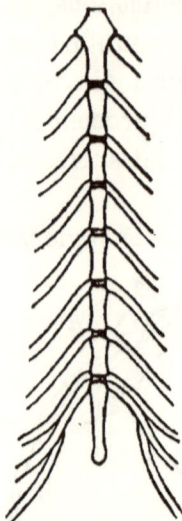

FIG. 56.—Sternum of dog, showing the separate elements of which it is composed.

In the sharks the cranium is a continuous box of cartilage, only perforated for the passage of nerves and blood-

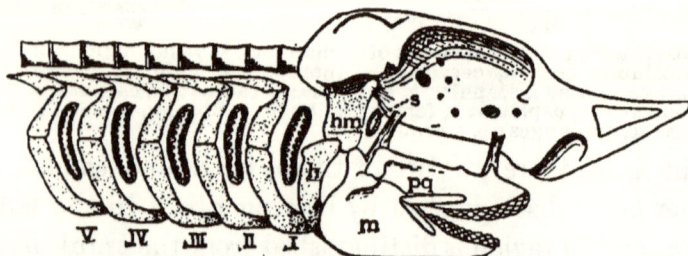

FIG. 57.—Skull and branchial arches of a shark. *h*, hyoid, and *hm*, hyomandibular form the suspensor of the lower jaw, *m* (Meckel's cartilage); *pq*, upper jaw (palato-quadrate) ; *s*, spiracle; I—V, gill-arches, between which are shown the gill-clefts.

vessels. In the other vertebrates some or all of this cartilage becomes replaced by bone, either by direct conversion (**ossification**) or by substitution. The bony cranium (unlike the cartilaginous cranium) is not a continuous wall, but is composed of separate bones firmly united together, the number varying between wide limits, being most numerous in the lower and reduced by fusion in the higher forms.

In the sharks the facial skeleton is very simple, being represented by the upper and lower jaws, and by a few cartilages supporting the lips. The upper jaw is not firmly united to the cranium, but is held in position by muscles and ligaments, while the lower jaw is hinged to the upper, and not to the cranium. Comparisons, which cannot be described here, show that the upper jaw of the shark is not the same as the upper jaw in the other vertebrates. In them numbers of other bones are added to the skull, and

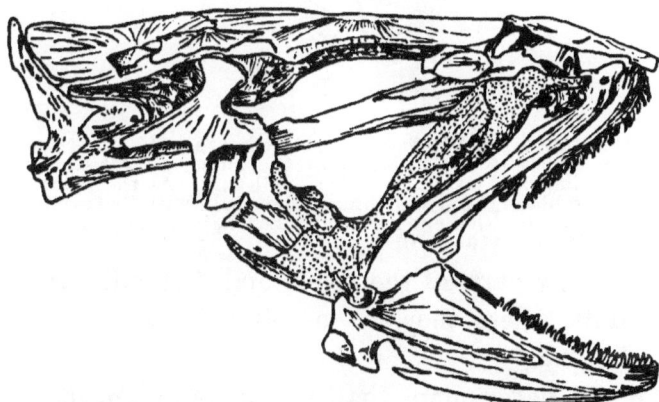

FIG. 58 —Skull of cod. (After Hertwig.) The dotted portion is the equivalent of the upper jaw of the shark (Fig. 57).

the upper jaw of the shark is only comparable to three pairs of bones, known to the anatomist as the palatines, pterygoids, and quadrates.

The visceral skeleton consists of bars of cartilage on either side of the throat between the gill-slits, the series being united below (Fig. 57). These **gill-bars** serve to keep this region, weakened by the openings, from collapse. The most

anterior of these gill-bars has the special name of **hyoid**.
There is some evidence tending to show that the lower jaw
and the palato-quadrate bar are but modified gill-bars.
With the disappearance of gills in the higher vertebrates
the branchial arches tend to disappear, and in birds and

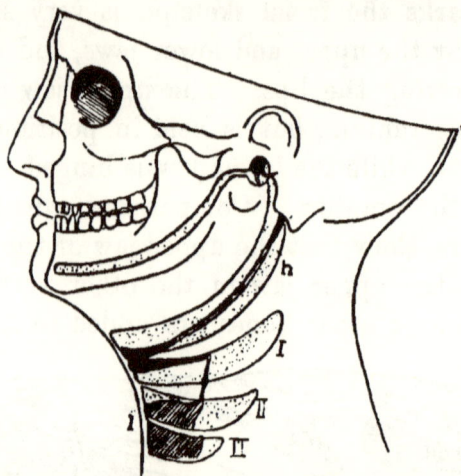

Fig. 59.—Diagram (after Wiedersheim) showing the relation of permanent
structures (dark) to the gill-bars of the embryo (dotted). *h*, hyoid
arch; *l*, cartilages of larynx; I, II, III, gill-bars. At the front of *h* and
I is shown in black the hyoid bone of the adult, with its two horns; behind the ear, at the other end of the hyoid arch, is (black) a piece (styloid
process) which joins the skull.

mammals only parts of the hyoid and first gill-bar remain
in the adult, where they are largely employed as supports
for the tongue.

There are never more than two pairs of appendages in
the vertebrates. These are the fore and hind limbs. In
their skeletons these are much alike, and in each can be
recognized arches of bone (**girdles**) uniting the limb to the
trunk, and the skeleton of the limb proper. These girdles
are known respectively as the shoulder or pectoral and the
pelvic girdle. In the fishes the girdles are simple arches,
and the skeleton of the limbs is largely composed of fin-rays
to support the flattened swimming organ.

In those vertebrates which support the weight of the body upon the limbs the appendicular skeleton is more complicated. In its typical condition the **shoulder-girdle** consists of three bones, which meet * to afford attachment for the skeleton of the fore limb. One of these bones, the shoulder-blade (**scapula**), is dorsal. It never joins the vertebræ, but is united to the trunk by muscles and ligaments.

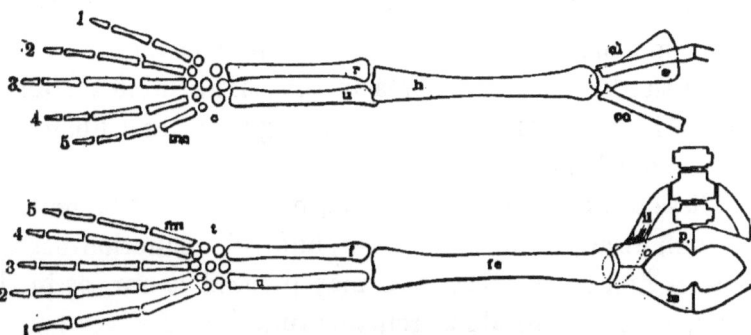

FIG. 60.—Diagram of fore and hind limbs of a terrestrial vertebrate, with one half of their girdles. *c*, carpus; *cl*, clavicle; *co*, coracoid; *f*, fibula; *e*, femur; *h*, humerus; *il*, ilium; *is*, ischium; *mc*, metacarpus; *mt*, metatarsus; *p*, pubis; *r*, radius; *s*, scapula; *t*, tarsus; *u* (in upper) ulna, (in lower) tibia; 1–5, digits, each composed of phalanges.

The other two extend ventrally from the shoulder-joint and meet the sternum. Of these the anterior is the collar-bone (**clavicle**), the posterior the **coracoid**.

In the **pelvic girdle** there are likewise three bones, which at their point of junction give rise to the hip-joint. The dorsal bone is the **ilium**, which articulates with the sacral vertebræ (p. 129), while below are found the **ischium** and **pubis**, the latter being the more anterior. Ischium and pubis unite with their fellows of the opposite side, thus completing the arch.

In the pelvic girdle the parts mentioned are pretty constant, but in the shoulder-girdle other bones may be added,

* The clavicle frequently does not enter into the formation of the shoulder-joint.

or either coracoid, or coracoid and clavicle may disappear. In the birds the clavicles unite, forming the wish-bone (**furcula** .

The bones of the fore limb (Fig. 60) are: a single bone (**humerus**) in the arm; two bones (**ulna** and **radius**) side by side, in the forearm; a series of nine bones (**carpals**) in the wrist; five longer bones (**metacarpals**) in the palm; and several rows (**phalanges**) of five bones in the digits. In the hind limb the conditions are closely similar: a single **femur** in the thigh, **tibia** and **fibula** in the shank, nine **tarsals** in the ankle, five **metatarsals** succeeding these, and finally the phalanges of the toes.

These are the typical numbers, but they may be reduced through disappearance or fusion, and this reduction usually appears first in the toes, and may proceed so far, as in the horse, that one toe alone remains functional.

The nervous system consists of a central and a peripheral portion, the latter consisting of nerves going from the central system to all parts of the body. To these should be added the organs of general and special sense.

The central system consists of an anterior **brain**, passing behind into the **spinal cord.** The brain is contained in the cranium, the spinal cord passes through the tube formed by the neural arches of the vertebræ.

The spinal cord is somewhat cylindrical, tapering behind,

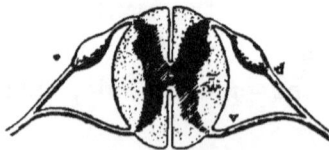

FIG. 61.—**Diagrammatic section** of spinal cord. *d*, dorsal nerve-root; *g*, gray matter; *v*, ventral nerve-root; *w*, white matter.

and contains in its centre a small canal. Nerves arise from the cord in pairs in regular sequence, and pass out between the vertebræ to all parts of the body and to the limbs. Each of these **spinal nerves** has two places of origin (roots) from the cord—one near the dorsal, the other near the ventral surface, but after a short course these roots unite into a common trunk. These roots differ in structure and function. The dorsal root bears a nervous enlargement or **ganglion**; the ventral has no such structure. Experiment shows that the dorsal root is concerned in bringing sensations to the central nervous system, and, if it be cut, the parts to which it goes will be without feeling. The ventral root, on the other hand, is motor; i.e., it controls the action of muscles, etc. If this root be cut, the muscles, glands, etc., which it supplies are paralyzed. Hence we may speak of the dorsal roots as **afferent**, since they bring sensations to the central nervous system; while the ventral roots are **efferent**, because they carry nervous impulses in the opposite direction.

The brain must be recognized as an enlarged and specialized portion of the central nervous system. The canal of the spinal cord continues into the brain, enlarging them into four cavities or **ventricles,** connected by narrower portions.

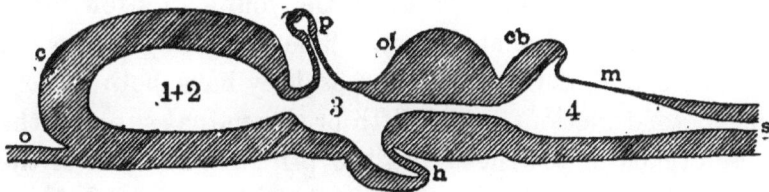

FIG. 62.—Diagram of vertebrate brain. *c,* cerebrum : *cb,* cerebellum; *h,* infundibulum ; *m,* medulla ; *o,* olfactory nerve ; *ol,* optic lobes ; *s,* spinal cord ; 1—4, ventricles.

In the brain five portions may be distinguished. Beginning in front, these are: (1) the **cerebrum,** composed of right and left halves or hemispheres, and containing in their interiors the first and second ventricles; (2) the smaller 'twixt-brain, with thin walls and enclosing the third ventricle; (3) the thick-walled **optic lobes**; (4) the **cerebellum** ; (5) the **medulla oblongata,** the fourth ventricle being

contained in cerebellum and medulla. In the lower vertebrates these five regions are nearly equal in size, but the
higher we go in the scale the larger proportionately do the
cerebrum and the cerebellum become, until in man the
cerebrum weighs about nine tenths of the whole brain.

From the brain are given off, typically, twelve. pairs of
nerves, which are spoken of both by numbers and by their
proper names. The majority of these are unlike the spinal
nerves in that they have but a single root, and are correspondingly either sensory or motor. Thus the first or **olfactory** nerve, which goes to the nose; the second or **optic**
nerve, to the eye; the eighth or auditory nerve, distributed
to the ear,—are purely sensory. On the other hand, the
third, fourth, and sixth (**oculomotor, trochlearis,** and
abducens) nerves go to the muscles of the eye; the eleventh*
(**accessorius**) goes to the muscles of the shoulder-girdle;
and the twelfth (**hypoglossal**) goes to the muscles of the
tongue. These nerves are purely motor, but it must be
remembered that the twelfth in the young of a few forms
has a dorsal ganglionated root. The remaining nerves are
like the spinal nerves in so far as they have both sensory
and motor functions. The fifth or **trigeminal** supplies the
sense organs of the head and the principal muscles of the
jaws. The seventh (facial) goes to the superficial facial
muscles, and in the lower vertebrates supplies certain sense
organs in the skin, but in man has lost its sensory functions.
The ninth (**glossopharyngeal**) goes to the tongue and
pharynx; while the tenth (**vagus or pneumogastric**) supplies the sense organs of the lateral line (p. 137) of the trunk
and sends branches to the stomach, lungs, gills, heart, etc.

Connected with the nervous system are the sense organs.

---

* This occurs in no ichthyopsidan vertebrate.

The skin contains small **touch organs** connected with afferent nerves, and these are for the recognition of pressure and temperature. Possibly allied to these are the organs of the **lateral line,** which are found only in the aquatic Ichthyopsida. These organs are sometimes free on the surface, sometimes in pits, while not infrequently the pits are connected by canals running beneath the surface, with openings to the exterior here and there. This line of organs is plainly seen on the side of the body in most fishes. On the head, however, it frequently branches greatly and becomes enormously extended in this way. The occurrence of these structures in aquatic forms only would suggest that their function is connected with that element; but what that function is, is not well understood.

The taste organs are within the mouth, principally on the tongue. They are poorly developed in some vertebrates, better in others.

The olfactory organs are always placed in front of the mouth. They consist of a membrane folded so as to expose a great amount of surface, and this surface is covered with the sense structure, connected with the ends of the olfactory nerve. In the fishes the sacs containing this membrane

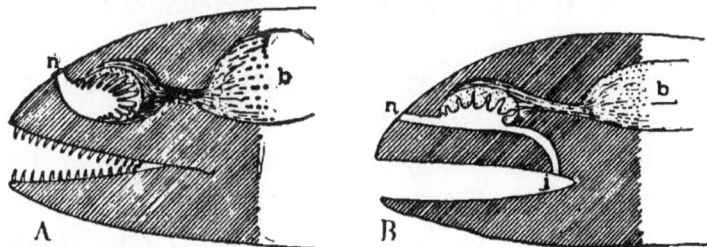

FIG. 63.—Relations of the olfactory organ, *A*, in fishes, *B*, in higher vertebrates. *b*, brain ; *i*, internal nostril ; *n*, external nostril. The sensory surface is folded.

have only external nostrils, but in all others they are placed at one side of a tube, which leads from the external nostril

to the back part of the mouth.   Hence a fish can perceive
odors in the water only as it swirls in and out of the nasal
sac.   In the air-breathing forms, odors in the air are drawn
with the breath over the sensory surface.

The essential part of the ear consists of a thin membranous sac on either side of the head.   In three places this
sac is so pinched as to form small tubes (**semicircular
canals**) open at either end into the main sac.   The whole is
filled with fluid in which are numerous minute solid particles (**otoliths**).   At one end of each tube and at places in
the sac are sensory organs connected with the auditory
nerve.   Sound-waves entering the ear set the fluid in motion,
causing the otoliths to strike the sensory organs and thus
to stimulate the nerve.

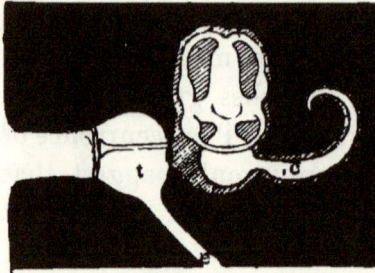

FIG. 64.—Diagram of mammalian ear. *c*, cochlea; *e*, Eustachian tube; *s*, semicircular canals. connected with the central sac and separated from the surrounding bone (black) by a space; *t*, tympanic cavity closed externally by membrane, and traversed by a bone, which conveys the sound-waves to the inner parts.

In the sharks this ear-sac is placed behind and medial to
the spiracle (p. 17).   In the higher vertebrates the spiracle
becomes closed on the outside, but the rest of the structure
remains, and is known as the **Eustachian tube,** and as its
outer end comes between the ear and the external world,
one or more bones usually extend across the tube to convey
the sound-waves to the sac.   In the frogs the outer end of
the Eustachian tube is closed by the large **tympanic
membrane** on the side of the neck.

In the higher vertebrates an external ear occurs.   This
consists of a tube leading inward to the tympanic membrane, and to this tube are frequently added structures to

catch and reflect the sound-waves into the tube. It should be mentioned that the ear is more than an organ of hearing; it is also an organ for maintaining the balance, for if the ear or the auditory nerve be injured the animal can no longer maintain its equilibrium.

The eye is built on the plan of a photographic camera. The essential parts are a **lens** which brings the rays of light to a focus on the **retina,** and means for causing the image on the retina to stimulate the optic nerve. To these are added various accessory structures for protection, for regulating the amount of the light, etc. In the lower forms eyelids are absent, but higher in the scale folds of flesh are developed which can close over the organ. Many animals have three of these eyelids, two working vertically, the third, the **nictitating membrane,** extending from the inner angle eye over the transparent cornea. This nictitating membrane occurs in the eye of man as a small fold (**semilunar fold**), which has entirely lost its primitive protective function.

Over the whole globe of the eye is a tough layer, the **sclerotic coat,** which is usually white, and which may be cartilaginous or may even have bone deposited in it, as in many reptiles and birds. In front this layer becomes perfectly transparent, and is there known as the **cornea.** Inside of the sclerotic is found a densely black layer (choroid), and still within this the transparent **retina,** the outer portion of which is imbedded in the choroid. In front the choroid is continued into the **iris,** a circular muscle with an aperture the **pupil,** in its centre. This iris, which is colored, regulates by its enlargement and contraction the amount of light which is admitted to the visual parts of the eye. Back of the iris and held in position by a circular muscle and ligament is the transparent **lens.** In front of this lens is a watery fluid (**aqueous humor**), while behind it and between it and the retina is the somewhat denser **vitreous humor.**

The optic nerve enters the eye from behind, passing through sclerotic, choroid, and retina, and is then distributed over the inner surface of the latter layer.

The eyeball is moved by six muscles, which are essentially

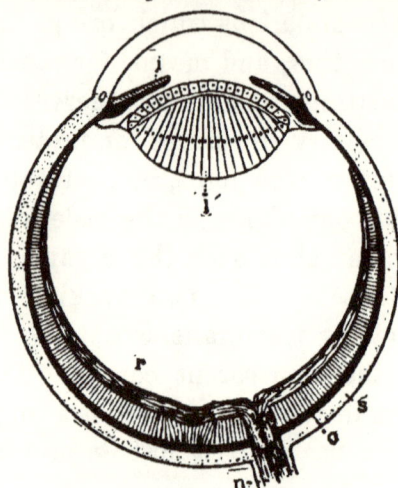

Fig. 65.—Diagram of vertebrate eye. *c,* choroid; *i,* iris; *l,* lens; *n,* optic nerve; *r,* retina; *s,* sclerotic.

alike in all vertebrates. Four of these are straight or **rectus** muscles, two are **oblique.** These muscles are controlled by the three eye-muscle nerves (p. 136).

The alimentary canal runs through the body from mouth to vent. In it several parts can be distinguished.

The mouth, at or near the anterior end, is without fleshy lips, except in the mammals. The mouth is frequently armed with teeth, and even in those groups, like the turtles and the birds, where they are absent the germs occur in the young, a fact which points to the descent of these from toothed ancestors.

The tongue is formed as a fold of the floor of the mouth, and is usually supported by a skeleton (hyoid, p. 132) derived

from the first or first and second visceral arches. In some
it is without powers of motion, but frequently it is very
mobile. Usually it is attached behind, the front margin.
being free, but in many batrachia it is attached in front
and folded back in the mouth.

The mouth-cavity is succeeded by the **pharynx**, a region
distinguished by containing the respiratory openings (internal nostrils, gill-slits, glottis, p. 93).

Behind the pharyngeal region is the digestive tract proper.
In some vertebrates it is scarcely possible to distinguish

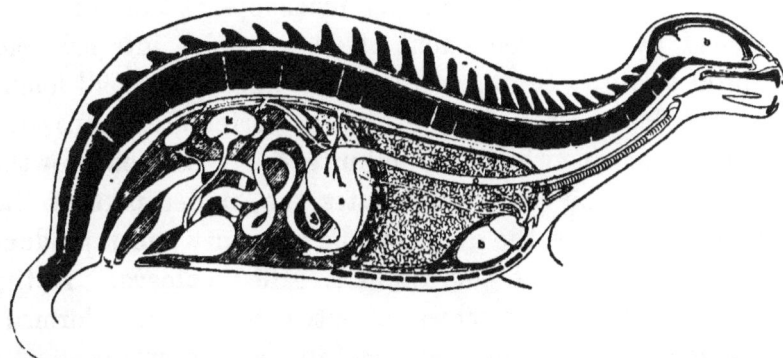

FIG. 66.—Diagram of the digestive tract of a mammal. *b*, brain; *d*,
diaphragm; *h*, heart; *i*, intestine; *k*, kidney; *l*, liver; *o*, œsophagus;
p, pancreas; *s*, stomach; *sp*, spleen; *v*, vent.

regions in it, but in most cases several distinct portions
occur. Those usually to be recognized are the following:

The pharynx communicates with the **gullet or œsophagus**, a muscular tube which frequently serves only to carry
food back to the stomach. On the other hand, a part of this
tube may be expanded into a glandular food-reservoir or
crop (birds).

In some fishes and batrachia the stomach is hardly differentiated from the œsophagus, but in other forms it is well
developed, with muscular and glandular walls. It may
even be divided into several portions. Thus in birds (Fig.

26) we frequently find two parts, one chiefly glandular while the other (gizzard) is extremely muscular. In the ruminants (p. 118) the specialization is carried farther, and we find four divisions to the organ.

The intestine is the absorptive portion of the alimentary canal. In some it is short and straight, in others long and convoluted, there being usually a correlation between length of intestine and the character of the food, this region being longer in the vegetable feeders. Increased absorptive surface is obtained in several ways, in addition to lengthening of the intestine. In the lower Ichthyopsida this is accomplished by the development of an extensive internal fold (spiral valve). In others there are numerous small longitudinal folds, while in the highest vertebrates transverse folds occur on which are minute finger-like outgrowths (villi). In the lower vertebrates the hinder part of the intestine receives the ducts of the excretory and reproductive organs, and at such times is called a **cloaca.** In the mammals, the monotremes excepted, no cloaca is formed. The vent is on the lower surface, in the median line.

There are several accessory structures connected with the alimentary canal. Thus frequently salivary glands are present, emptying into the mouth. Behind the stomach the ducts of the liver and pancreas pour in their secretions, while in many fishes well-developed **pyloric cæca** occur, just behind the stomach, which have a digestive function.

The digestive organs are supported in the body-cavity by a thin membrane (**mesentery**) which bears blood-vessels, etc., and which is attached to the dorsal wall of the body-cavity. This mesentery in reality is but the continuation of the lining (**peritonæum**) of the body-cavity.

Vertebrates respire in three ways: by gills, by lungs, and by the skin. Gills arise first as outpushings or pouches in

the sides of the pharynx, and then these break through to
the exterior, giving rise to gill-slits or clefts, through which
water taken in at the mouth can pass out.   On the sides of
these clefts the gills proper are developed.   These are thin-
walled leaves or filaments with a rich blood-supply, and
through these thin walls there is an exchange of dissolved
gases (oxygen and carbon dioxide) between the water and
the blood.

In the septa between the gill-slits are the gill-bars or cartilages (p. 131); and from the septa there grow out, in the

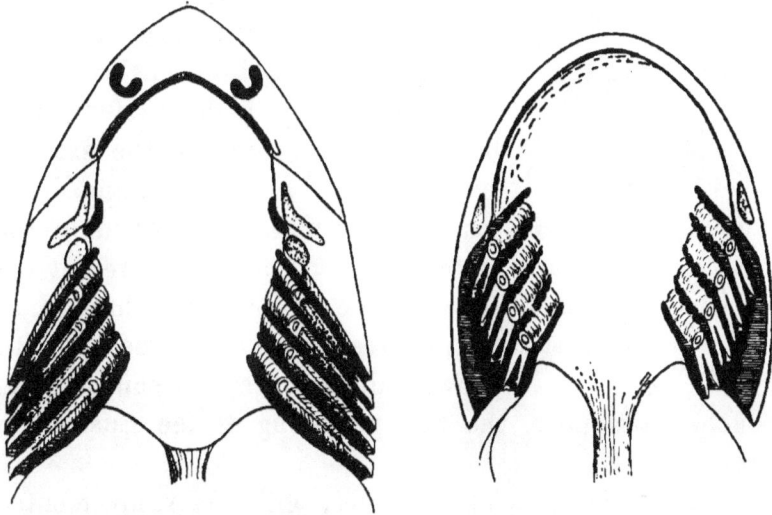

Fig. 67.—Relations of gills, gill-openings, etc., in a shark (left) and a
teleost (right).

larval batrachia, fleshy fringes, the external gills.   In most
batrachia these external gills are later absorbed and replaced
by internal gills, which in turn may disappear upon the
assumption of an aerial respiration.

The number of these clefts varies between four and eight,
but in all the anterior cleft has largely lost its respiratory
function.   In the sharks it becomes modified into the

spiracular cleft; in the higher vertebrates it enters into the structure of the ear, giving rise to the cavity of the drum and to the Eustachian-tube.

In the sharks each cleft opens separately to the exterior; but in ganoids and teleosts the hyoid septum gives rise to a fold (**operculum**) or " gill-cover," which grows back over the external openings, so that there is apparently but a single slit externally. A little con-

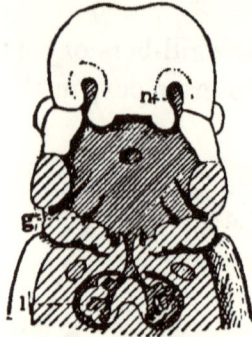

FIG. 68.—**Human embryo** (after Hertwig), with the floor of mouth and and throat removed, to show the rudimentary gill-slits. *g. l,* lung; *n,* nostril, still connected with the mouth.

sideration will show that there is little real modification. In the anurous batrachia a similar fold is found, but this unites again with the body-wall behind the gills, thus enclosing the external openings in an **atrium**, with but a single opening to the exterior (p. 50). In the sauropsida and mammals gill pouches are formed in the embryo, but according to recent observers these never break through, so that no real clefts are formed. With growth all but the first pair of these pouches disappear, the first persisting as the Eustachian tube.

In all vertebrates above fishes, gills are supplemented (batrachia) or replaced by lungs. These are paired sacs richly supplied with blood-vessels, and connected with the external world by means of a tube (windpipe or **trachea**) which opens by the **glottis** upon the floor of the pharynx. The trachea is usually strengthened by the development of cartilages in its wall, some of which may become large, as in the case of the human " Adam's apple." The lungs themselves may be simple sacs, but usually they become greatly folded, thus increasing the respiratory surface. In the batrachia, which lack diaphragm and ribs, air is forced into the lungs by swallowing; in the reptiles and birds it is drawn in by means of the muscles (**intercostals**) between the ribs; in the mammals the intercostals are reinforced by a transverse muscle (**diaphragm**), Fig. 66, which crosses the body-cavity.

In the ganoids and bony fishes exists a structure, the **swim-bladder** or **air-bladder**, which is usually thought to represent the lungs. In the lower teleosts (Physostomi) it is connected with the alimentary canal by a duct opening on the *dorsal* wall of the pharynx, but in others (Physoclisti) this duct closes long before the adult condition is reached. In the lung-fishes, on the other hand, the structure is double and its duct ventral.

Connected with the respiratory system are two glands of problematical function. One of these, the **thyroid**, is formed from the floor of the pharynx. The other (the **thymus**) arises from the gill-pouches, and in the higher vertebrates disappears in adult life. In the calf it forms the "neck sweetbread." Both these glands are without ducts, and the part they play is obscure.

In the circulatory system three parts may be recognized: (1) a central propelling organ, the heart; (2) arteries, carrying the blood away from the heart; and (3) veins bringing it back. Between arteries and veins are interposed minute tubes, the **capillaries.**

The heart is a muscular organ, enclosed in a special sac of the body-cavity, the pericardium. In the heart can always be distinguished a receptive portion (**auricle**), which receives the blood as it comes from the veins, and passes it on to the true propelling organ, the **ventricle.** This latter has strong muscular walls, and when it contracts, the blood, prevented by a valve from returning to the auricle, is forced out through the artery (**ventral aorta**) connected with the ventricle.

In all fishes there is but a single auricle and a single ventricle, but when lungs appear, as in the batrachia, the auricle becomes divided, and now one half (the right) receives the blood from the body, while the left auricle takes the blood returning from the lungs. These both pour the blood into the single ventricle. In the reptiles we find the beginning of a division of the ventricle, which becomes complete in the crocodiles and continues in birds and mammals. In these forms the left auricle pours its blood into the left ventricle, while the same relations exist between the auricle and ventricle of the right side.

In the fishes the blood leaves the ventricle by an arterial trunk, in which, when best developed, we can distinguish a **conus** with valves inside to prevent the blood flowing back into the ventricle; or a **bulbus**, without valves, and in front of these the **ventral aorta**. From this lateral vessels (**afferent branchial arteries**) are given off, and these pass up through the branchial septa. Consequently the number of these arteries primarily depends upon the number of gill-clefts. In the septa the arteries break up into capillaries which pass through the gills, and collect in **efferent branchial arteries** which pass above the pharynx. Here they unite and give rise to the main trunk, the dorsal aorta, which runs, above the alimentary canal, through the body, giving off vessels to all parts.

From these vessels the blood passes through the capillaries and is collected in veins which bring it back to the heart to repeat the circuit. In this circulation the blood changes in its character. When it enters the heart it bears nourishment obtained from the alimentary canal, and waste from all parts of the body. Its color is a dark purplish red. In its passage through the gills it rids itself of one kind of waste (carbon dioxide) and absorbs oxygen from the water. This exchange is accompanied by a change of color to bright red. The other waste is gotten rid of in the kidneys. In

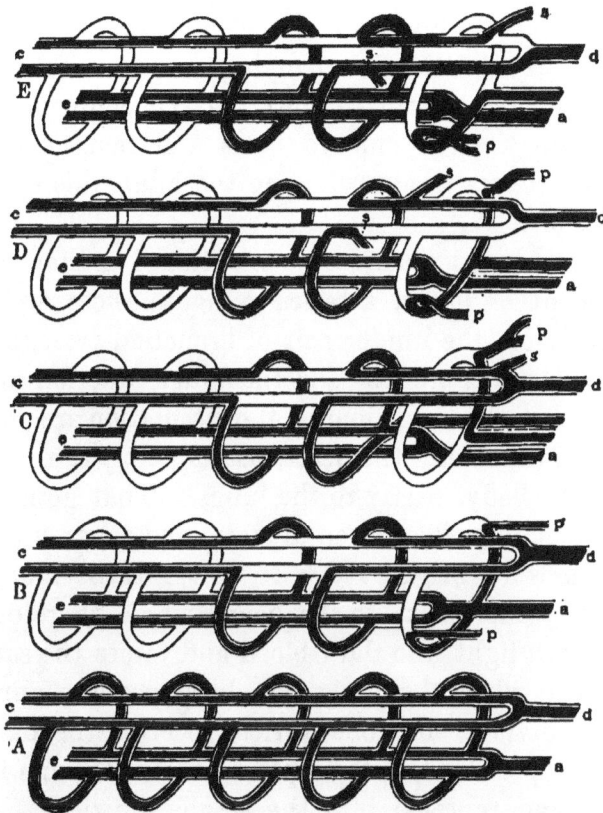

Fig. 69.—Diagram of the arterial arches and their modificaitons in various vertebrates. *A*, fish; *B*, batrachia; *C*, snake; *D*, bird; *E*, mammal. *a*, ventral aorta; *c*, internal carotid; *d*, dorsal aorta; *e*, external carotid; *p*, pulmonary artery; *s*, subclavian. Drawn by Dr. F. D. Lambert.

the capillaries of the body it gives up its oxygen and nourishment to the surrounding parts, and becomes loaded anew with carbon dioxide and other waste, changing color again to the dark red. From this account it will be seen that in the fish only blood charged with impurities passes through the heart.

From the arrangement of blood-vessels found in the fishes (sharks) all the conditions found in the higher vertebrates may be derived, simply by enlarging some vessels and suppressing others. Some of the changes involved may be made out from the accompanying diagrams (Fig. 69) in comparison with your dissections, the explanatory statement being made that in embryo birds and mammals *paired* branchial arteries occur, while in the adult this symmetry is largely lost.

One point particularly to be mentioned is that with the development of lungs, arteries going to these organs are developed from the hinder pair of branchial arteries.

When the gills are lost and the lungs function as respiratory organs, the conditions of the circulation are changed. The blood, in leaving the heart, passes partly to the various parts of the body, partly to the lungs. That going to the latter organ loses its carbon dioxide, and takes up oxygen and changes to bright red. It now returns along with blood from other parts to the heart, which therefore now receives both light and dark blood and forces the same out again. But when the lungs are developed the auricle of the heart divides, and one auricle receives the dark, the other the light blood, both emptying their contents in turn (in frogs and reptiles) into the single ventricle. It was therefore formerly thought that the blood sent out through the ventral aorta must necessarily be mixed; but this is not the case. By means of a peculiar valve the red blood is sent to the body, the dark blood to the lungs.

As has already been mentioned, in crocodiles, birds, and mammals the ventricle is also divided, and now one half of the heart contains only bright, the other only dark, blood. The division is also carried farther, for the last arch (going to the lungs) becomes connected with the half of the heart which receives the dark blood, while the rest of the arches are similarly related to the other half of the heart.

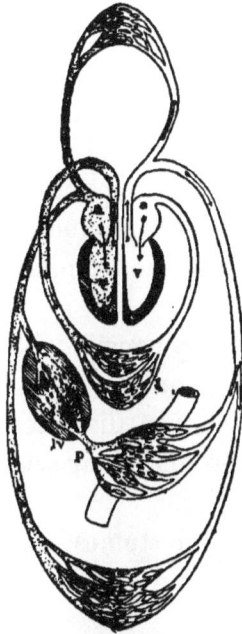

The blood itself should have a moment's attention. It consists of a fluid (**plasma**) in which float myriads of minute solid bodies (**corpuscles**). The plasma is a pale yellow in color, the red of the blood being due to certain of the corpuscles, which are therefore known as the red corpuscles. Other corpuscles are colorless, and are called white corpuscles or **leucocytes**. The red corpuscles carry the oxygen and carbon dioxide, the plasma the nourishment and the other waste. The plasma is further peculiar in that when withdrawn from the veins it soon solidifies or "clots."

FIG. 70.—Diagram of the circulation in a mammal. The arrows show the direction of the flow; the vessels carrying red blood are shown white, those carrying dark blood, shaded. *a*, auricles; *l*, lung; *lv*, liver; *p*, portal vein bringing the blood from the intestine; *v*, ventricle.

The excretory organs (**kidneys** or **nephridia**) are very complicated structures. In a few words, they may be described as a pair of organs lying in the dorsal wall of the body-cavity close to the median line. Each kidney is richly supplied with blood, and it extracts from this fluid the nitrogenous waste and pours it into an excretory or urinary duct which empties behind, near the anus.

The reproductive system is closely related to the excretory organs. In all except a few fishes the sexes are separate. In the females, eggs are formed in special structures, the ovaries, and when ripe the eggs are passed out to the exterior by means of a tube (oviduct) developed from the urinary duct. This passage may be rapid, or the egg may remain for a time in the oviduct and there undergo its development, as is the case in certain members of all groups of vertebrates except birds.

In the male, corresponding to the ovaries in position, etc., are the testes, which produce the male reproductive element, which is also carried off by a part of the primitive excretory duct.

All vertebrates produce eggs, but these vary considerably in size. In the mammals the diameter is about $\frac{1}{100}$ of an inch, the ostrich lays an egg about 5 inches in diameter, while the egg of one of the extinct birds of Madagascar was equal in size to 150 hen's eggs.

The Vertebrates are divided into Cyclostomes and Gnathostomes.

CYCLOSTOMATA.

The Cyclostomes include a few eel-like forms, commonly known as lampreys and hagfishes. These differ from the other Vertebrates in many points, some of which are mentioned here. Bone is entirely lacking, and cartilage is feebly developed. Vertebræ are scarcely recognizable, and there are no traces of paired fins, although dorsal and caudal fins may occur. The mouth, as the name Cyclostome implies, is circular, but is incapable of closure like that of other vertebrates, since movable jaws are lacking. Inside of the mouth are horny teeth (few in the hagfishes, many in the lampreys), but these are chiefly used for holding, not for biting or crushing. The tongue is very large.

There is but a single nostril on top of the head. The

gills are placed not in simple slits, but in large pouches on the sides of the neck, and these pouches may either open separately to the exterior or by means of a tube which leads to a single opening. The number of gill pouches ranges between six and fourteen on either side.

The Cyclostomes are subdivided into two groups, accordingly as the nostril communicates with the throat or not. As examples of the first, the hagfishes may be cited. These are all marine, and are capable of secreting a large amount of mucus from their bodies, so that a few hagfish in a pail will convert the water into a jelly-like mass. These fishes are parasites, and work their way into various fishes, like the cod, and when once inside they eat up all the flesh and viscera, leaving nothing except the skin and bones.

The second group is represented by the lampreys. Some of these are marine, others live in fresh water, while many

FIG. 71.—Lamprey (*Petromyzon marinus*). After Goode.

of the marine forms ascend streams in spring to lay their eggs. By means of their circular mouths, horny teeth, and sucking tongues, the lampreys attach themselves to fishes, from which they suck the mucus and frequently the blood. In some places the large sea-lampreys are regarded as delicacies, but usually they are not esteemed as food.

GNATHOSTOMATA.

In the Gnathostomata jaws are always developed; the skeleton, whether of cartilage or bone, is a true support to the body; usually paired limbs occur, and the nostrils are always paired. The general account of the Vertebrata in the preceding pages applies especially to the Gnathostomes. This group is divided into three subbranches:

SUBBRANCH   I.—ICHTHYOPSIDA (p. 55).
SUBBRANCH II.—SAUROPSIDA (p. 85).
SUBBRANCH III.—MAMMALIA (p. 97).

CHORDATA.

There are a few characters of the Vertebrates which are shared by other forms. These features are: (1) the possession of gill-slits ; (2) a nervous system which is entirely on one side of the alimentary canal; and (3) a notochord which lies between the alimentary canal and the nervous system. The existence of this notochord has given the name Chordata to the assemblage. There are four divisions or branches of the Chordata, only three of which need mention here.

BRANCH I.—LEPTOCARDII (Lancelets).

The few species of lancelets (*Amphioxus*) are all marine, and occur in warmer seas. They have a body which is fish-like, but they differ from all fishes in the absence of a true heart and of a skull. The gill-slits are numerous

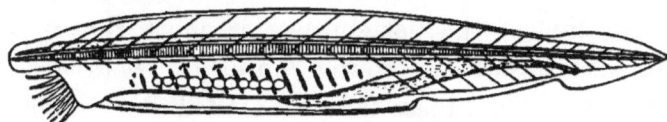

FIG. 72.—Diagram of Amphioxus (after Hertwig and Boveri). Above (dotted) is the nervous system ; below it (cross-lined), the notochord ; the mouth is surrounded by the circle of tentacles ; below the notochord is the region of gill-slits ; the vent is near the posterior (right) end below.

(about sixty), and these empty into a gill-chamber recalling in some features that of the tadpoles. The notochord runs the whole length of the body, and a stomach is lacking, the liver emptying into the intestine just behind the gills. Limbs or paired fins are absent, but there is a median fin at the end of the body. The animals are about two or three inches long, are almost perfectly transparent, and bury themselves in the sand, only the mouth end, encircled by a fringe of delicate filaments, appearing above the surface. They are without any economic importance, but their extremely simple structure makes them intensely interesting to the naturalist.

BRANCH II.—TUNICATA.

The fact that these forms had any relationship to the Vertebrates would never have been suspected had one studied only the adults. · When, however, the development was studied, it was perceived that these forms had larvæ in which there was a notochord, gill-slits, and a nervous system much like that of the Vertebrates; in short, that in shape and in structure these young Tunicates were decidedly tadpole-like. Then these tadpoles settled down upon some object and passed through a metamorphosis in which the tail was lost, the nervous system was contracted into a mass, and the body became more or less saccular and covered with an external envelope or "tunic," which gives the name to the group.

Of these Tunicates there are many varieties, but the essential features of the adult can be made out from the generalized figure given. The body is globular, and shows on the outside two openings. One of these is the mouth, which communicates with a gill-region perforated by numerous gill-slits. At the bottom of this pharyngeal region is the œsophagus, which leads to stomach and intestine, the latter twisting so as to terminate at the bottom of a cloacal chamber, which opens to the exterior by the other aperture mentioned. The water, which passes through the gill-slits, is collected, and passes into the same cloacal chamber. The nervous system consists of a centre or ganglion between the two openings, from which nerves radiate to the various parts. There is a heart at the

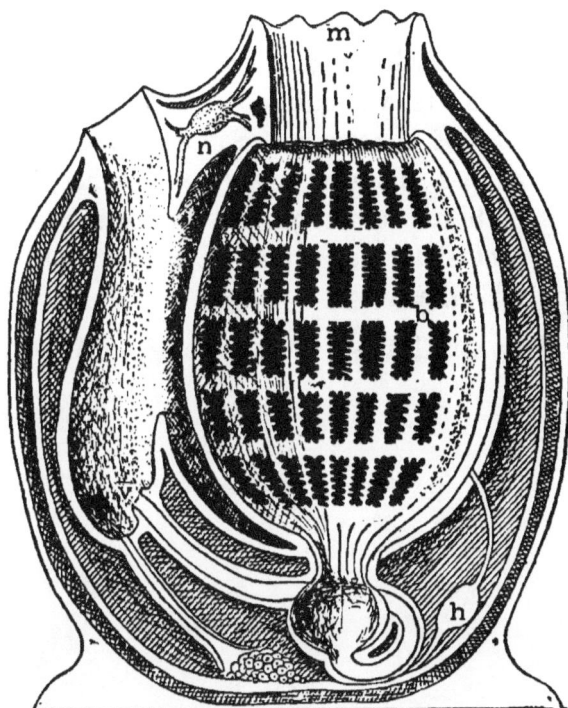

Fig. 73.—Diagram of a Tunicate. *b*, branchial chamber, perforated by gill-clefts, and connecting at the bottom with the œsophagus which leads to the globular stomach, and thence by the intestine to the vent, *v; h*, heart; *n*, nervous system; *m*, mouth.

opposite side of the body, and a peculiarity of this organ is that it regularly changes in its action, the blood flowing in a direction opposite to that which it followed a moment before.

The species of Tunicates are numerous, and show great variety of form. A characteristic of many is the power to reproduce by budding, and as a result there are formed large colonies, the members of which are more or less intimately connected with each other. In some cases the animals resulting from budding produce eggs, and these eggs grow into forms unlike their parents but like those from which the parents were budded. In other words, the child does not resemble the parents, but the grandparents. This peculiarity is called "alternation of generations."

The tunicates are all marine, and they abound in the seas of all parts of the world. Some of them are known from their shapes and color as "sea-peaches," others as "sea-pears," while a common name for all is "sea-squirts," due to the fact that they squirt water from the openings upon being disturbed.

BRANCH III.—VERTEBRATA (p. 127).

Each pupil will require at least two specimens. One of these should be opened along the back, as described below, and placed for some days in alcohol in order that the internal parts may become hardened, thus better fitting them for dissection.

Can you distinguish two regions in the body? How many joints (**segments** or **somites**) can you distinguish in the posterior region or **abdomen**? Can you see segments in the anterior region (**cephalothorax**)?

Examine a segment (the third) of the abdomen. How is it joined to the segments in front and behind? Are the parts between the segments as hard as the walls of the segment? What is gained by this arrangement? How does the wall of the segment differ from a ring? To what part of the ring are the appendages (**swimmerets**) attached? How many of these are there on the segment? In a swimmeret make out the basal joint (**basiopodite**), having two leaf-like branches, one towards the median line of the body (**endopodite**), the other outside of this (**exopodite**). Draw the segment and appendages from in front.

Compare the segments behind the third with that one. Do all have the two-branched appendages? How are the swimmerets of the sixth segment modified? How does the last segment (**telson**) differ from the others? Where is the vent? Compare the appendages of the first and second

157

abdominal somites with those of the third. In the male they are peculiarly modified. What numerical relations do you find between somites and appendages in the abdomen? (**Savigny's law**).

Examine the lower surface of the cephalothorax, and see if you can find traces of segments, especially in the region near the abdomen. How many appendages on one of these somites? How many pairs of large legs, including the "pincers," do you find? In the hinder pair of legs how many joints do you find? Can you distinguish exopodite and endopodite? Compare this leg, joint by joint, with the big claw. What change would make it into a pinching-organ? How many of these legs are furnished with pincers? Look on the inside of the basal joints of the legs for openings (outlets of the reproductive organs). If they occur on the middle pair the specimen is a female; if on the last pair it is a male. What is the sex of your specimen?

Study the appendages (**mouth-parts**) in front of the big claws. In order to do this properly it will be necessary to remove those of one side one by one, by grasping the base of the appendage with the forceps and pulling it out. Be careful to get all of each appendage, and nothing else. The three hindermost (or outer) mouth-parts are the jaw-feet (**maxillipeds**). Compare the hinder pair with the third swimmeret. Do you find basiopodite, exopodite, and endopodite? Compare it with one of the walking-legs. Which part, exopodite or endopodite, is lacking in the latter? Draw each of the maxillipeds.

In front of the maxillipeds come two pairs of accessory jaws (**maxillæ**). Remove them carefully, and draw. Look on the hinder maxilla for a large expansion, the **gill-bailer**. Removing these parts exposes the mouth, on

either side of which is a strong jaw (**mandible**).   How do
these jaws move in comparison with those of man ?   Take
one out, and see of how many joints it is composed.

The cephalothorax is covered above by a large continuous plate, the **carapax**.*   Does this show signs of segments.   With the forceps lift the hinder corner of the
carapax on the side where the mouth-parts still remain,
and see where it joins the body.   Then with the scissors
cut away the free portion, thus laying open the **gill-
chamber**, and exposing the body-wall and the numerous
gills or **branchiæ**.   Are any of these attached to the legs or
to the body-wall?   Move the maxillæ, and see the operation of the gill-bailer.   Can water obtain free access to the
gills ?

In front of the mouth occur the "feelers" or **antennæ**.
Could these be compared to the legs ?   Can you find exopodite or endopodite in them ?   Examine the basal joint
of the larger or posterior one (the antenna proper), and
find an opening, the outlet of the green gland (see below).
Is it in any way comparable in position to the reproductive
opening ?   In the smaller feelers (antennulæ) look for the
ear-sac on the upper surface of the basal joint.   (It is
covered with a thin membrane, around which hairs are
arranged.)   Above the antennulæ are the **eyes**.   Are they
movable ?   Examine the black portion (**cornea**), and see
the small portions (facets) of which it is composed.

Make a tabular arrangement of the appendages of the
body,† and ascertain by Savigny's law (p. 158) how many
segments there are in the body of the crayfish.   Compare
the segments, and see how their diversity is brought about

---

* Often written carapace.   The spelling here adopted is preferable.

† For reasons which cannot be discussed here, the eyes are not regarded as appendages comparable to the others.

by under-development (**atrophy**) of one part and **over-**development (**hypertrophy**) of another. (The carapax is really but the dorsal portions of the antennal and mandibular somites, the line crossing its middle being the line of union of these two.)

Make a side view of the crayfish, twice the natural size, naming the parts.

Internal Structure.

The dissection should be made under water, the specimen, back upwards, being held in position by being pinned to the wax bottom of the dissecting-pan, the pins passing through the telson and large claws. Open the crayfish along the back by cutting away the carapax with the scissors, taking care not to injure the underlying parts. Continue the cuts backward, removing the upper surface of the abdomen.

Just beneath the carapax, behind the impressed line crossing it, is the oblong whitish heart. How many openings through its walls can you find? How many tubes (**arteries**) leading from it? With the forceps gently tip the heart to the side. Can you find more openings or more arteries? Is there a chamber (pericardium) around the heart? Trace the arteries as far as you can without injuring other parts.

Beneath the heart, and projecting from beneath it, are the paired reproductive organs. Do those of the two sides connect? Can you find the ducts leading down from them? Where do they end? Still farther in front is the large thin-walled stomach, and on either side of this, and reaching back to the heart, is the liver, reddish in the crayfish, green in the lobster. Tip the stomach backwards and see the **œsophagus** or tube leading to it from the mouth. Tip it forwards and find the **intestine**. Can you find the ducts leading from the liver to the intestine?

Draw the viscera, etc., as far as made out, adding the intestine later.

Cut away heart, liver, reproductive organs, and trace the intestine to the vent. Is it the same size throughout?

Take out the stomach, being very careful not to injure other structures when cutting the œsophagus. Open the stomach and find the teeth; how many? Try to see how the teeth grind the food.

In the front part of the body, close to the antennæ, find the **green glands** (paired). Their openings have already been found. They are excretory in function (kidneys).

Cut away the (white) muscles in the abdomen, being careful as you approach the floor, and expose the hinder part of the central nervous system (**ventral cord**). How are the enlargements (**ganglia**) arranged with reference to the segments of the abdomen? Are the nerves given off from the ganglia, or from the cord (**commissure**) connecting them? Trace the ventral cord forward into the cephalo-thorax, carefully breaking away the hard parts which cover it, and follow it forward to the **brain**, in front of the mouth. How many ganglia do you find in the cephalo-thorax? Do any show signs of being double? Is the commissural cord single or double in this region? Is there a ring of the nervous system around the œsophagus? Can any of the nervous system be said to be above, or any below, the alimentary canal? From what part do the nerves to the antennæ and eyes arise?

Draw the nervous system from above.

Can you make out three regions in the body: head, thorax, and abdomen? Where would you draw the lines between the regions?

Examine a thoracic segment. Does it resemble in any way an abdominal segment of a crayfish? Study the legs. Can you find exopodite and endopodite? How many legs do you find? Are any of them terminated with pincers? Look beneath the thorax for thin overlapping membranes attached to the bases of the legs. They will be found only in females. Between them and the lower surface of the body is a chamber or **brood-pouch** to contain the eggs or young. Do you find anything in this cavity?

How many segments do you find in the abdomen? Notice the last pair of abdominal appendages extending behind the body. Turn the animal on its back, and with the needle pull apart the flattened plates on the lower surface of the abdomen. These are the gills. Do they present any of the characteristics of appendages? How many of these gills do you find? Examine them all and see which ones bear white spots (air-chambers). Draw a pair of these gills.

Examine the head. Where are the eyes? Are they on stalks? What are the peculiarities of the antennæ? Can you, with the lens, find another pair of minute antennæ? The mouth-parts form a short, thick projection

162

beneath the head.   Pick this apart with the needle.   How
many pairs* of mouth-parts can you find?   Counting all
the appendages of the head, how many segments should
there be in this region?

COMPARISONS.

With one column for crayfish, the other for sow-bug,
give answers to these questions:

(1) Are head and thorax united?

(2) Are the eyes on movable stalks?

(3) How many pairs of walking-feet, counting the
pincers as such?

(4) Where are the gills?

(5) Are both exopodites and endopodites present?

---

* The two of the hinder pair are united, but should be counted as
a pair.

DECAPODA.

Those forms which are commonly known as crayfish, shrimps, lobsters, prawns, and crabs are collectively known as Decapods, from the fact that, including the large claws, they have ten walking-feet. Besides this we find that they all have eyes on movable stalks, the anterior part of the body (thirteen segments) is covered by a fold of the integument known as the carapax, and the gills are (usually) borne packed away in a gill-chamber above the walking-legs.

This group of Decapoda is subdivided into three divisions, or "suborders," according, among other things, to the characters presented by the abdomen. In the MACRURA it is, as shown in the crayfish, very large, and is carried well extended; in the BRACHYURA it is much smaller, not nearly so large as the cephalothorax, and is folded up beneath the latter region so that it is not visible from above. In the third group, the ANOMURA, the abdomen is intermediate between the conditions found in the other groups, and frequently it is much softer than the other regions.

Of the Macrura the most important are the lobsters, which are large marine forms differing in few points, except size, from the fresh-water crayfish. These play a great part in the food-supply of northern Europe and the eastern United States. They are mostly captured by sinking large wooden traps (lobster-pots) baited with refuse fish, and at intervals hauling up the pots. The number thus

taken upon tne shores of New England and Canada amounts
to octween twenty and thirty
million annually. Crayfish
are used largely as food in
Europe, and are bred in ponds
for the market, but in America they are largely neglected.
Shrimps and prawns are
largely salt-water forms, but
some of the prawns occur in
fresh water in the warmer
parts of the world. The line
between the two is not easily
drawn except by saying that
the body of the shrimp
is flattened (depressed) from
above downwards, while that
of the prawn is compressed
(flattened from side to side).
In America, "shrimp salad"
is almost universally made
from prawns.

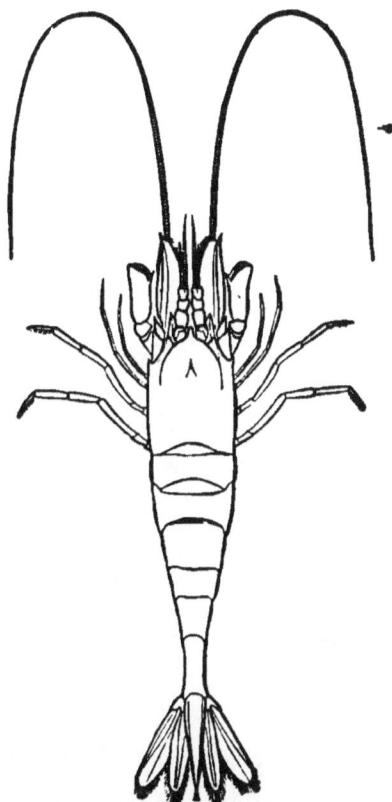

Of the Anomura, the most
interesting are the so-called

FIG. 74.—Common shrimp (*Crangon vulgaris*). From Emerton.

hermit-crabs. These are somewhat lobster-like, but the
abdomen is but slightly hardened, and so, to protect this
vulnerable part of the body, the crab inserts it in a deserted
snail-shell, and this "house" he carries about with him
wherever he goes, retreating into it and closing the opening
at the approach of danger with his solid pincing-claws.
With increase in size the crab must move into a larger
shell. In other Anomura the back is soft, and these

"false hermits" carry half a clam-shell about with them to cover their weak point.

FIG. 75.—Hermit-crab (*Eupagurus bernhardus*) in a snail-shell.   From Emerton.

Only a few of the true crabs or Brachyura live in fresh water.   In the tropical and semi-tropical regions are those

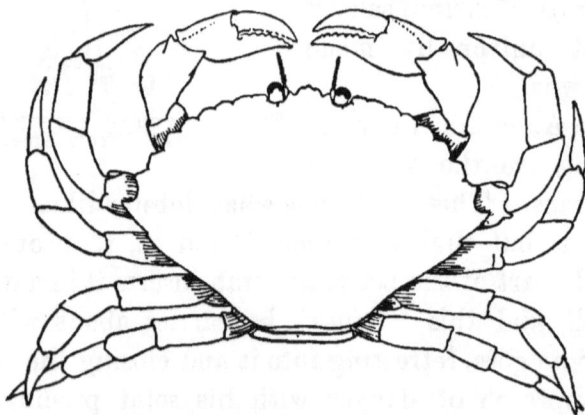

FIG. 76.—Shore-crab (*Cancer irroratus*).

which live on the land; but the great majority—a thousand different kinds—live in the sea.   The larger species have some economic value as food, but all of them are important as scavengers. In America "soft-shelled crabs" are prominent in our markets at the proper season of the year. During the rest of the time this crab, known as the "blue crab," has as hard a shell as any crab, but when the proper moment comes the shell splits across the hinder margin, and out from this opening comes the body covered only with the thinnest skin, and at this time alone is it a "soft shell." All other crabs molt or shed their skin in the same way, the new skin rapidly growing hard again, but the blue crab is the only one taken in sufficient abundance at this time to be of economic importance.

TETRADECAPODA.

Contrasted to the Decapods are the fourteen-footed or Tetradecapodous forms, of which the sow-bug is one type. In these we can distinguish clearly head, thorax, and abdomen, the joints of the thorax being freely movable on each other. The eyes are not placed upon movable stalks, but are scarcely elevated above the general surface of the head. Most of these forms are marine; a few live in fresh water, and still fewer, like the sow-bugs and pill-bugs, upon the land. All are small, those which reach two inches in length being the veritable giants among the group.

There are two subdivisions of Tetradecapods: Isopoda and Amphipoda.

In the Isopods the body is depressed, as in the sow-bug, and the gills are borne under the abdomen. Most of the Isopoda feed upon decaying matter, but some have become parasites upon other animals, and have consequently so changed their appearance that one knowing only the adult would never regard them as Isopods at all. But the young settle the question, since before they begin their parasitic life they are regular Isopods.

In the Amphipods the body is compressed from side to side, and the gills are borne on the thoracic region between the legs. These forms are familiar to all visitors to the shore under the common name of "beach-fleas," a name which those forms living under dried seaweed, etc., have won for themselves through their leaping powers. Others

168

live in the ocean itself. None of them have any economic
importance aside from their acting as scavengers and
serving as food for fishes.

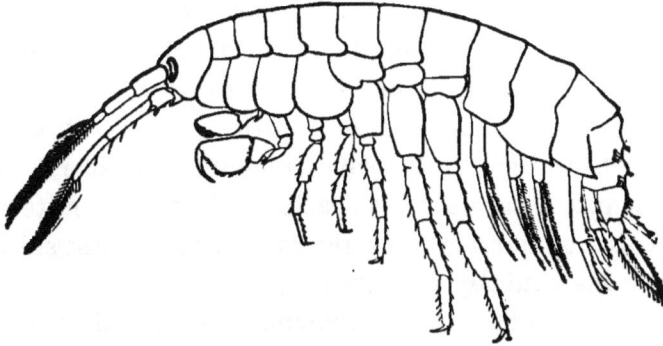

FIG. 77.—Beach-flea (*Gammarus ornatus*). From Smith.

Can you distinguish three regions—head, thorax, and abdomen—in the body? Where would you draw the lines between these regions? and why at these points?

Notice that the abdomen is made up of a series of rings (**segments or somites**) essentially like each other. Examine a ring at about the middle of the abdomen, and see that it is made up of dorsal and ventral hardened halves, united by a more flexible membranous portion. Look at the side of the somite and find a small opening (**spiracle**). How many somites bear similar spiracles? Has any somite more than a pair of spiracles? Could you speak of these spiracles as being **segmentally arranged?**

Examine the base of the abdomen and see that its first segment is incomplete. Look at the lower surface and see if you can find the lower half of this ring. On the sides of this first ring notice a large oval thin spot (**tympanic membrane**), the so-called ear. Can you find a spiracle near the ear?

The tip of the abdomen varies in shape in the two sexes. In the female it is provided with two pairs of pointed outgrowths (**ovipositor**). The male lacks these, and the tip is rounded and frequently upturned. Study this region carefully in each sex, making out the following points:

In the male notice that the ventral halves of the terminal segments are much larger than the dorsal portions. (This overgrowth is called **hypertrophy.**) Counting from the base, how many rings can you find in the whole abdomen?

170

Are any except the first incomplete? Lift the parts on the dorsal side of the tip of the abdomen and find the vent. On the dorsal side between the vent and the tenth somite is a broad plate (**supra-anal plate**), and on either side of this is a small outgrowth from the tenth segment (**anal cercus**). Are these anal cerci movable? Could they be regarded as jointed appendages? To which somite do they belong?

In the female study the terminal somites in the same way as in the male. Do you find the same dorsal and ventral halves? Are any of them hypertrophied? Do you find vent and anal cerci? Examine the ovipositor.* Are its parts movable? See if they are attached to the eighth and ninth segments.

Draw side and dorsal views of male and female abdomens, making each sketch at least five inches long. Insert all features made out, lettering everything.

In the thorax recognize three segments: **prothorax, mesothorax, metathorax,** the first overlapping the others something like a cape. How many legs are attached to the prothorax? Look in the membrane joining the pro- to the mesothorax for a spiracle. Study a prothoracic leg. It is made up of a series of joints. Joining the leg to the body are two short joints (**coxa** and **trochanter**), then comes a long **femur**, next an almost equally long **tibia**, and lastly, a several-jointed foot or **tarsus.** Notice how freely the head moves upon the prothorax by means of a flexible

---

* As its name implies, the ovipositor is of use in laying the eggs. By means of it the grasshopper bores a hole in the earth, and then the packets of eggs, passing down through the tube formed by the four members of the ovipositor, are deposited in the ground. Other allied species use the ovipositor for placing the eggs in leaf-buds or in the stems of certain plants.

" neck." Separate the prothorax from the head and from the mesothorax, and draw it from the side.

Study meso- and metathorax together. Notice that on the back the line between these somites is very distinct; trace this line upon the side, and thence to the ventral surface. Do you notice any other lines which seem to divide meso- and metathorax ? Can you trace them on all surfaces ? Do you find any spiracles in this region ? How are the legs related to the somites ? Can you recognize in each the same parts found in the prothoracic legs ? Where are the wings ? Are they alike ? What is the prevailing direction of the ribs or " veins " in them ? Can either pair be folded like a fan ? Is there anything to protect the hinder pair when at rest ?

Draw a side view of meso- and metathorax, inserting expanded wings, legs, etc.

Remembering what was found out about the crayfish and sow-bug, and considering the thorax and tip of the abdomen of the grasshopper, do you find anywhere a segment bearing more than a pair of jointed appendages ? * So far as your present knowledge goes, would you be justified in saying that a pair of jointed appendages indicates a somite of the body ?   **(Savigny's law.)**

Notice that the head is made up of a large solid piece (**epicranium**), to which are attached various movable portions. On either side of the head is a large **compound eye**. With a sharp knife slice off one of these eyes and examine it under a low power of the microscope. Why is it called compound ? What is the shape of the parts (**facets**) of which it is composed ?

Look on the front of the head for the smaller bead-like

---

* For reasons which cannot be discussed here, the wings of grasshoppers, etc., are not considered as jointed appendages.

simple eyes or **ocelli**.   How many of these do you find, and how are they arranged ?

On the front of the head, below the eyes, is a broad fold, the **clypeus**, to which is attached a movable upper lip (**labrum**) covering the mouth in front.   Near the eyes arise two long, slender feelers or **antennæ**.   Could they be re-garded as jointed appendages ?

On the lower side of the head is the **mouth**, surrounded by a series of appendages or **mouth-parts**.   Beginning be-hind, remove these one after another with forceps and needle.   The most posterior is the lower lip or **labium**. It is in reality double, and consists of the united basal joints and, arising from these on either side, a several-jointed **palpus**.   Draw the labium × 10, and then take off and draw the pair of appendages, the **maxillæ**, next in front.   Notice that in these the basal joints are enlarged, one forming a sharp cutting-organ, the other a more fleshy portion to hold the food in position.   The terminal parts form a **palpus**, somewhat similar to the labial palpus.   Still further in front come the jaws or **mandibles**.   Move these with the forceps.   Do they work in the same way that your own jaws do ?   Draw them, and then draw front and side views of the head, labelling all the parts.

Have you found any traces of segments in the head? How many pairs of jointed appendages have you found? According to Savigny's law, how many segments* must there be ?

INTERNAL STRUCTURE.

The internal structure of the grasshopper in its larger features is readily made out.   Select a large female for the

---

* Study of the embryos of some insects makes it probable that there is one more segment in the head than is shown by Savigny's law.

purpose of dissection; pin it out, back uppermost, in the dissecting-pan, in water just deep enough to cover it, and with fine scissors cut away the dorsal wall of the abdomen, taking great pains to remove nothing but the hard parts. In spite of all care the beginner will probably remove the **heart**—a delicate tube lying along the middle of the back— with the dorsal wall.   Continue the cuts forward, removing the dorsal wall of the thorax. Notice the large muscles which move the wings. If the specimen has been freshly killed, the most striking feature will be a series of silvery-appearing air-tubes, **tracheæ**, which connect with the spiracles and ramify all parts of the body. In alcoholic "hoppers" these are distinguishable only with difficulty. Between the body-wall and the viscera will be found the light-colored **fat-body.**

In the anterior part of the abdomen, on either side, is a cluster of long oval yellow eggs, and from each mass of eggs a delicate tube (**oviduct**) may be traced backwards to the region of the ovipositor. Separate the masses of eggs and find, between and below them, the dark-colored alimentary canal. Follow this forward and back and make out in it the following parts: In the hinder half of the abdomen the **intestine**, which in front passes into the much larger **stomach.** At the junction of the stomach and intestine are a number of fine tubes (**Malpighian-tubes**) which are excretory in function. At the anterior end of the stomach are a number of larger double-cone shaped tubes, the **gastric cæca**, and in front of these is the large brown **crop.** The crop is connected with the mouth by a narrow tube, the gullet or **œsophagus.**

Remove the alimentary canal by cutting through œsophagus (close to the crop) and intestine, and look upon the floor of the abdomen for the nervous system. Can you find enlargements **(ganglia)** in this? How are they arranged with regard to the somites? Follow the nervous system forward, if possible, into the head. Can you find cords passing around the œsophagus as in the crayfish? Is there a brain above the gullet? Does the alimentary canal pass through the nervous system?

Draw a diagram of a section passing through the thorax, showing the body-wall, wings, legs, spiracles, egg-masses, nervous cord, alimentary canal, and heart in their relative positions.

THE CRICKET : LABORATORY WORK.

Do you find the same regions as in the grasshopper?
Are there the same number of segments in the abdomen?
and in the thorax? Are the wings and the feet the same
in number in the two forms? In the place of the cerci
what do you find? Could you call these jointed appendages? How many parts do you find in the ovipositor of
the female? What changes in the grasshopper ovipositor
would be necessary to make it like that of the cricket?
Can you split any of the parts of the ovipositor of the
cricket? Can you find the ear?

In the head are there the same eyes, antennæ, and mouth-
parts? Do the mandibles work in the same way? Look
on the second long joint (tibia) of the foreleg for the ear.

176

"JUNE-BUG" (BEETLE) : LABORATORY WORK.

How does the size of the head compare with that of the grasshopper? Can you find both ocelli and compound eyes? Notice the antennæ on the front of the head. Draw one. What changes would you need to make in the antenna of a grasshopper to make it like that of the June-bug? Can you find labrum, mandibles, maxilla, and labium as in the grasshopper?

How many pairs of walking-legs do you find? Do you find segments to correspond? What name must be given to the large segment just back of the head? Examine a leg : Do you find in it the same segments that occur in the leg of the grasshopper?

Lift one of the hard outer wings (elytra). Do you find veins, like those of the grasshopper, in the elytron? Is there a second pair of wings? Are they as long as the elytra? How are they folded?

Study the abdomen. Can you find the membranous portion uniting dorsal and ventral halves of the somites? Are spiracles present? Can you find any "ears"? How many segments can you count in the abdomen? Do you find anal cerci, or ovipositor? Separate the flaps at the hinder end of the abdomen. Can you find any additional segments? Draw a beetle from above with elytra and wings extended.

177

DRAGON-FLY: LABORATORY WORK.

Which pair of wings are the larger? What is the general arrangement of the veins in the wings? Is the head freely movable? What is the size of the compound eyes? How many simple eyes do you find? How would you describe the antennæ?

Are the mouth-parts fitted for biting? Do they move like those of a grasshopper? Do you find upper lip (labrum)? maxillæ? labium? What is the character of the mandibles? Are they toothed? Have any of the mouth-parts palpi?

Do you find all three of the thoracic segments? Are those present of equal size? Are they firmly united to each other? What is the relative size of the legs? How many joints in the foot?

How many segments do you find in the abdomen? Are any of them partially divided? On what ones do you find spiracles? Are there appendages on any of the abdominal segments?

What peculiarities do you find in the antennal joints? Are both compound eyes and ocelli present? Are the mandibles, like those of the grasshopper, fitted for biting? How do the other mouth-parts compare in shape with those of the grasshopper? Do you find a "tongue"?

How many thoracic segments? Are all freely movable? Which is the smallest? Which bear wings? Which pair of wings is the largest? Are the veins of the wings many or few? Are the wings transparent?

Does the abdomen join the thorax by its whole width, as in the grasshopper? or is there a slender stalk joining the two? How many abdominal segments do you find? Squeeze the abdomen and look for the sting. Does it compare in any way with the ovipositor of other insects? Where was it before pressure was applied?

COMPARISONS.

Rule a sheet of paper with columns for Grasshopper, Beetle, Dragon-fly, and Wasp, and write the answers in each to the following questions:

(1) Are ocelli present?

(2) Are the antennal joints equal in size?

(3) Are the maxillæ and labium short and stout, or long and slender?

(4) Are any of the thoracic rings free?

180   *ELEMENTS OF COMPARATIVE ZOOLOGY.*

(5) Which thoracic ring is the largest?

(6) Which pair of wings is the larger?

(7) Describe the general structure of the fore wings.

(8) How many segments in the abdomen?

(9) Are any appendages besides those of the ovipositor present on the abdomen?

(10) With which column should the cricket be placed?

The name Orthoptera, which is given to the group containing the grasshoppers, crickets, locusts, cockroaches, etc., means straight-winged, and alludes to the general course of the veins of the wings of most forms. This is, however, not a feature of great importance, for indeed we find species which are absolutely lacking in wings, but which are, in other respects, so closely related to the grasshoppers that they too must be included in the Orthoptera. When we take all of these Orthopterous forms we see that they agree in a number of points, some of which may be mentioned. The jaws are strong and fitted for biting hard substances; the antennæ are usually long and thread-like; ocelli are always present; the prothorax moves freely on the mesothorax; the abdomen is ten-jointed, and it usually bears on its tenth somite movable cerci; the ovipositor is large and cannot be withdrawn into the abdomen; the anterior wings serve as covers for the second pair, and these last are folded longitudinally, when at rest, like a fan.

Besides these points, which should have been made out by the student, there is another feature not readily discovered in the classroom. The young Orthopteran hatches from the egg with all the legs and segments of the adult, which it resembles much in general appearance, except in the following particulars: it is smaller in size, with a disproportionately large head, and it lacks the wings characteristic of the full-grown form. It is most voracious, and with

much eating increases rapidly in size. But since it is
enclosed in a hard outer wall, incapable of growth, it has
frequently to cast off this non-elastic "skin" and to grow
a new one, larger than the old. This **molting** is accomplished by a splitting of the old skin down the back, and
from this hole the animal draws itself, and now, its skin
being soft, it can readily increase in size. Gradually, however, the skin becomes thicker and harder, and the process
of molting must be repeated. With each of these molts
the animal grows more like the adult, the wings appearing

FIG. 78.—Young grasshopper with the wings just beginning to appear.
After Emerton.

first as small pads upon the back, and with later molts
attaining the final size. It is an easy matter to follow
these changes by catching the young hoppers in the spring,
and keeping them in a breeding-cage, feeding them frequently with fresh grass and leaves. The student must
keep this history in mind when studying the peculiarities
of the beetles.

With few exceptions the Orthoptera are injurious to
human interests, since they are vegetable-feeders, and, as
they often occur in immense numbers, they can destroy all
crops throughout large districts.

Possibly the most disagreeable members of the group are
the cockroaches, flattened forms, many of them wingless,

which are familiar from the persistence with which they haunt our dwellings, etc., after they have once been introduced. Our familiar " Croton bug " is an immigrant from Europe, but we have also our native species. Insect-powder and eternal vigilance are the only means known to rid a building of these pests.

Strangest of our Orthoptera are the "walking-sticks"; long, wingless animals which feed upon the oak and which, as they stand motionless upon a twig, can scarcely be distinguished from the twigs themselves. The species figured is foreign.

Grasshoppers and locusts are much alike, and are usually confused by most people. Both are leaping forms, but the locusts have short antennæ and short ovipositors, while the grasshoppers have these parts long. The katydid is a grasshopper, while the "grasshopper" which in 1873–76 did such damage in our Western States is a locust. Closely allied are the crickets, whose ceaseless chirp is so monotonous upon summer nights.

Fig. 79. — Walking-stick (*Acanthoderus*). From Hertwig.

These make their song by rubbing their wing-covers together, and it is interesting that only the male can make the noise. The "ear" of the cricket is not upon the abdomen, but upon the fore legs. It is not certain that any of these structures are really for hearing.

COLEOPTERA (Beetles).

The beetles are all grouped under a common head of Coleoptera, the name of which means sheath-wings. Of beetles there are known over a hundred thousand different kinds, but all these agree in the following points: The mouth-parts are fitted for biting; ocelli rarely occur; the prothorax is large; the anterior wings* are converted into thick, horny wing-covers or elytra, beneath which are folded the much larger hinder wings.

From the egg of the beetle there hatches out a somewhat worm-like form popularly known as a "grub." This **larva**, as it is called, bears but the slightest resemblance to its parents. It eats and grows, without essentially altering its appearance until at last it undergoes a molt which results in a sudden change in its appearance. It is no longer worm-like, but looks more like the adult beetle. This stage, which is called the **pupa**, does not eat, but lies quiet in some cavity; after a longer or shorter period of rest it molts again and emerges the perfect beetle, after which, no matter how long it may live, it undergoes no further changes, nor does it increase at all in size. Forms which, like the beetles, pass through these abrupt changes are said to undergo a **complete metamorphosis.**

The beetles are divided into two great groups. In the one (Rhynchophora) that part of the head which bears the

* The elytra of beetles are apparently not the same organs as the anterior wings of grasshoppers or butterflies, but the distinction between the two cannot be made clear here.

mouth is prolonged into a snout; in the other there is no such prolongation. These are called the normal Coleoptera.

Of the normal Coleoptera some are beneficial to man, since they feed upon other insects. Here may be enumer- ated the brilliant tiger-beetles and the caterpillar-hunters, the habits of which have given them their common names. They are all extremely active. The water-beetles should be placed in the same category, for they and their larvæ feed upon the insects of our streams and ponds, and do not a little towards keeping the mosquitoes within bounds.

Another large group of beetles have the antennæ ending in a club or knob. Some of these, like the carrion-beetles, are of value, since they lay their eggs in decaying flesh, where the larvæ live and flourish, converting what otherwise would be a nuisance into another crop of beetles. Others, like the "ladybugs," are predaceous, feeding upon the smaller insects; but still others are unmitigated nuisances, since they have a taste for dried animal matter. Among these are the bacon-beetle and the far better known "buffalo-bug," which plays havoc with our silks and woolens, our carpets, and the specimens in our museums. In this same group belong the rove-beetles, forms in which the wing-covers are very short, not covering half of the long abdomen. Disturb one and notice the threatening way it moves its abdomen about, as if to sting. It is, however, perfectly harmless.

The spring-beetles and the fireflies agree in having the antennæ toothed something like a saw. The spring-beetles receive their common name from the fact that when laid upon their backs they will suddenly throw the body into the air. When opportunity offers, study the actions of one of these and see how the spring is arranged. Some of these spring-beetles are serious pests, for their larvæ are the well-known wireworms. The fireflies are interesting from their phosphorescent powers. Underneath the abdomen are the light-giving spots. Much attention has been given to this light-producing apparatus in the hopes of obtaining a solution of the problem of producing light without heat.

A large number of beetles have the terminal portion of the antennæ, like that of the June-bug, with a club formed of leaf-like joints. These are known as Scarabæans, from the sacred beetle (*Scarabæus*) of the Egyptians which belongs to the group. These sacred forms are represented in our country by the tumble-bugs, which lay their eggs in balls of manure which they trundle along the road until they find a suitable place to bury them. From the similar habits of the Scarabæus the Egyptians worked out quite a symbolism. "The ball which the beetles were supposed to roll from sunrise to sunset represented the earth; the beetle itself personified the sun, because of the sharp projections on its head, which extend out like rays of light; while the thirty segments of its six tarsi represented the days of the month." Other members of the Scarabæans, like our June-bugs, are vegetarians and do no little damage. As larvæ they feed upon the roots of the grass and other plants; as adults they devour the foliage. In the tropics occur Scarabæans of enormous size, some having bodies six inches in length.

The long-horn beetles live as larvæ in the solid portions of trees and shrubs, where they bore long tubes. The species usually have long antennæ, and many of them are beautifully colored. Structurally much like these borers are the shorter and more oval leaf-beetles, which do so much damage. Here belong the cucumber-beetles, the Colorado potato-beetle, and others which feed upon the grape, the asparagus, etc.; and near them are the so-called weevils which attack peas and beans.

The oil-bottles and blister-beetles are a curious group, since in their young stages many of them are parasitic upon other insects, while when adults they contain a peculiar substance which will raise a blister upon human flesh. Hence some of these are killed, dried, and form a regular article of commerce under the name of Spanish flies.

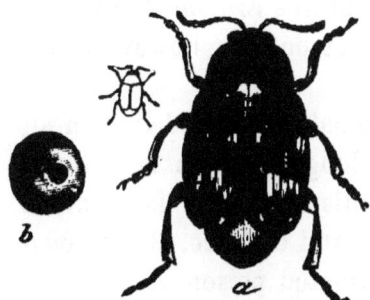

FIG. 80.—Pea-weevil (*Bruchus pisi*), natural size and enlarged. *b*, pea containing a weevil.

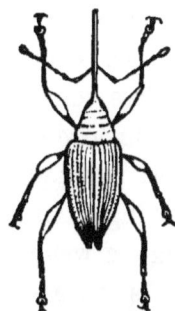

FIG. 81.—Hazelnut-weevil (*Balaninus nasicus*.)

The snout beetles (Rhynchophora) or true weevils are all injurious, since as larvæ and adults they feed upon vegetation. Some attack fruits, some eat grain, and others nuts. Certain ones burrow between the bark and solid woods of trees, excavating curious mines, while others bore into the solid wood.

HYMENOPTERA (Bees, Wasps, Ants.)

Bees, wasps, and ants are the better known representatives of this group, all the members of which agree in having four membranous wings (the front pair the larger) with comparatively few cross-veins. The mouth-parts are fitted both for biting and for sucking. There is a complete metamorphosis. So far as we can judge, these are the most intelligent of all insects, and the student who investigates their habits is continually rewarded by new facts, which show that their small brains are most highly developed. In other points of structure, however, they are much less complicated.

In the lower forms the female is provided with an ovipositor, frequently of great length, which is well adapted for boring. In the higher this ovipositor is modified into a sting—a weapon of offence and defence, the efficiency of which is increased by an associated poison-gland.

The lowest forms are the sawflies, the larvæ of which are vegetable-feeders, some eating the leaves of plants, others boring in the solid wood. A little higher in the scale come the gall-flies, those forms which lay their eggs in various plants and in some way so stimulate the vegetable tissue that strange growths—galls—are formed. Allied to these last are the ichneumon-flies, which lay their eggs in other insects. Here the larvæ hatch out, feed upon the host, at last destroying it. Then pupation comes, and the perfect insect emerges to repeat the process. Naturally these

188

ichneumon-flies are an important agent in keeping down injurious insects.

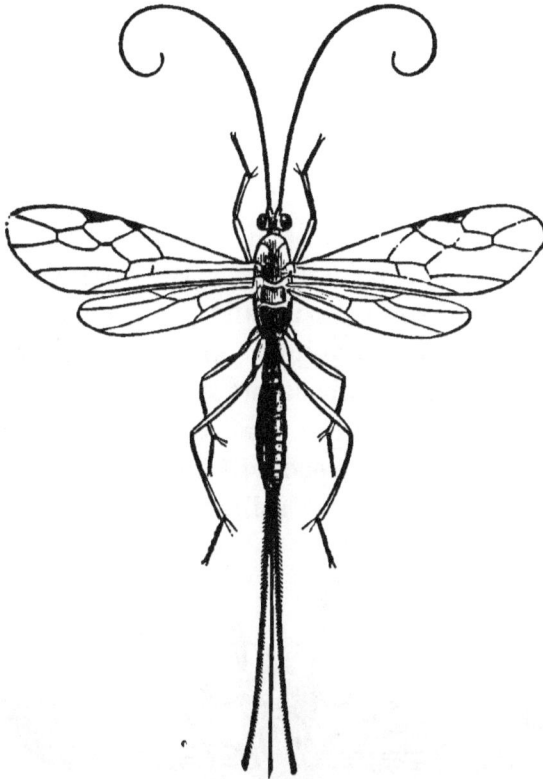

Fig. 90.—Ichneumon-fly, enlarged. From Riley.

The ants are possibly the most interesting of all insects. They are true communists. In them, as in the white ants (p.   ), there is a differentiation of the individuals into males, females, and workers, the latter being wingless. Any adequate treatment of these forms would of itself demand a book larger than this volume. The males and females take "wedding-flights," after which the male soon dies, while the females bite off their wings and henceforth have nothing to do except to lay eggs. These eggs are cared for by the workers, which, as the name implies, perform all the labor of the colony. They obtain the food, take care of the immature insects, build the nests, and carry on the wars. In their battles some ants always take prisoners, and these are kept as slaves. Some species of ants have depended on slaves so long that they are only able to fight, while did the slaves not feed them they would starve. No group of insects will better reward careful study than these.

The digger-wasps make mines in the earth or in wood in which they lay their eggs, usually placing with the eggs a supply of food for the young. Some use as food pollen and nectar of plants, while others store up insects or spiders which have been so stung that they are paralyzed, not killed. In this way the food will keep for a long time.

The true wasps are some solitary, some colonial, and in

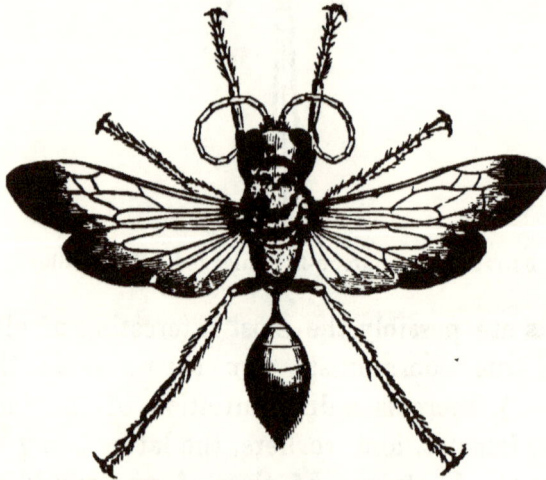

FIG. 91.—Sand-wasp (*Sphex*).

the colonial forms we find again, as in the ants, males, females, and workers, the workers being winged. Most of these true wasps (and hornets are wasps) build nests usually of half-decayed wood, which they chew into a kind of paper. Inside are the cells in which the eggs are placed and in which the young undergo their metamorphosis. Males and workers die in the autumn, but the females live through the winter and start new colonies in the spring.

Among the bees the honey-bees occupy the first place from their value as honey-storers. Indeed, so great is their value that hundreds of books and dozens of journals have been published dealing wholly with them. In each colony there are males (drones), females (queens), and workers, the latter imperfectly developed females. Soon all the drones are killed, and all the queens except one, and her sole duty is to lay eggs. If the queen be lost, the workers can take a larva that would otherwise develop into a worker, and by different food convert it into a queen. Wax is a secretion of the bee, honey is the nectar obtained by the bees from flowers, while the bee-bread is the pollen of flowers.

SQUASH-BUG: LABORATORY WORK.

Can you distinguish three regions in the body? How many legs do you find? Have these the same joints as in the grasshopper? How many joints in the tarsus? Do you find compound eyes, ocelli, and antennæ on the head? Examine the lower surface of the head and find the beak. See if with needles you can separate it in several needle-like parts. This can only be done with great care in so small a form as the squash-bug. The student, if successful, will find four needle-like pieces (mandibles and maxillæ) sheathed in a groove-like labium. Could this beak be used for biting and chewing, or for piercing and sucking? Notice the wings, drawing one of each pair. In the anterior wing are the basal and distal parts equally thick? How many joints in the abdomen? Where are the spiracles?

BUTTERFLY: LABORATORY WORK.

Notice the relative size of the wings. How are they carried when the insect is at rest? Rub the wings and notice your fingers. Scrape a wing with a scalpel and study the "dust" under the microscope. With what is the body covered? Sketch the veins in the wings.

Study the head. Below, notice the proboscis. Straighten it out with a needle. At the sides of the base of the proboscis see the labial palpi. Are the antennæ of the same size throughout their length? Sketch the head in outline, then remove the hairs, etc., which cover it and look for eyes and ocelli, inserting what you find in the drawing.

Do you find in the legs the same joints as in the legs of grasshoppers? Have the tarsi of all the legs the same number of joints? Is the base of the abdomen as broad as the thorax? Do you find cerci or ovipositor?

COMPARISONS.

With columns for Squash-bug and Butterfly, answer the following questions:

(1) What is the relative size of the two pairs of wings?
(2) How are they carried when at rest?
(3) Are they naked or covered with scales?
(4) Are either pair thickened at the base?
(5) Are distinct labial palpi present?
(6) Can the proboscis be used as a piercing-organ?

The Hemiptera are the true bugs. This term is frequently loosely applied, but any true bug has the following characteristics: Its mouth-parts are not fitted for biting,

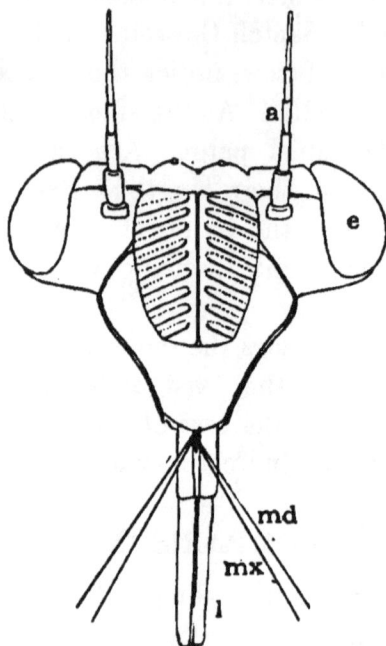

FIG. 82.—Head of seventeen-year locust to show the mouth-parts, etc. *a*, antennæ; *e*, compound eye; *l*, labium; *md*, mandible; *mx*, maxilla.

but for piercing and sucking. They are prolonged into a beak, consisting of a fleshy grooved sheath (labium) with four needle-like bristles (mandibles, maxillæ) in the groove. This organ is used for making holes in plants or flesh, and

194

also serves as a tube through which the bug sucks up the juices found. The bugs have an incomplete metamorphosis (p. 182), hatching from the egg much in the adult condition, except that wings are lacking.

Almost all the Hemiptera, when adult, have four wings, though there are a number of wingless forms. These wings are built upon two distinct patterns, and this serves as a means of subdividing the Hemiptera into two groups. In the one (HETEROPTERA) the basal half of the anterior pair of wings is thickened while the rest is membranous, and the wings themselves are held in an overlapping manner upon the back when at rest. This condition is familiar in the squash-bug. In the other group (HOMOPTERA) there is no such thickening of the basal portion of the first pair of wings, and these organs, when at rest, are placed upon the sides of the abdomen.

While most of the bugs are injurious to human interests, there are some which are a benefit to man, since they feed on injurious insects; and still others, like the cochineal- and lac-bugs, produce substances of value to man.

Of the Heteropterous forms some are aquatic, and of these the water-skaters, gliding over the surface of still water, are familiar to all. Others live most of their lives beneath the surface, and some of the larger of these water bugs can kill small fish, sticking the beak into them and sucking their blood.

Of the terrestrial forms none is more widely known than the bedbug, a form which is famed for its attacks on man. It is one of the bugs which never develop wings. From the pecuniary standpoint the chinch-bug is more important, since it attacks fields of grain, doing sometimes millions of dollars of damage in a single year. The young attack first the roots and underground stems, and later the stems themselves, killing them before they have had time to ripen the grain.

The squash-bug, which does such damage to pumpkin- and squash-vines, is another form of Heteropteran, as are those familiar forms which have no other common name than "stink-bug." No one who has ever taken into his mouth a berry over which one of these animals has travelled can

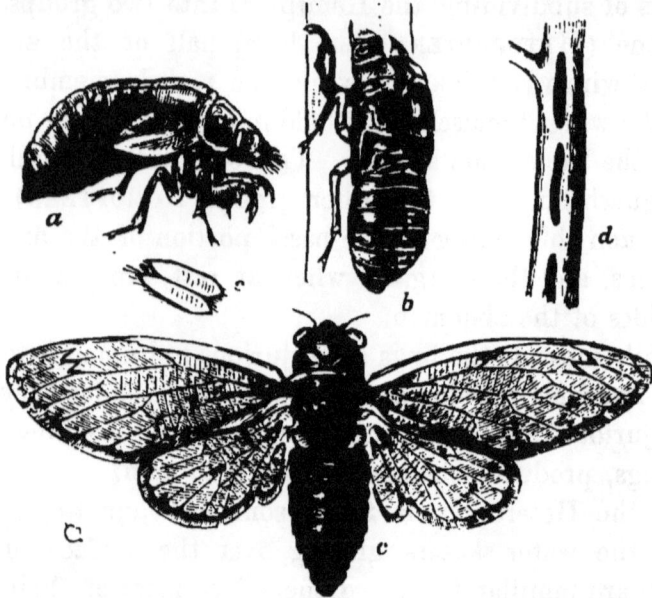

FIG. 83.—Seventeen-year locust (*Cicada septendecim*). From Riley. *a*, pupa; *b*, pupa-case from which the adult, *c*, has escaped ; *d*, twig bored for the deposition of eggs.

doubt the appropriateness of the name. However, these bugs are not alone in their malodorous qualities; many others, like the squash-bug and bedbug, also secrete a strong-smelling fluid, which of course protects them from birds and other insect-eating animals.

Among the Homopteran forms the cicadas come first. One of these, the "dog-day locust" (it is not a locust at all), is familiar from its shrill note heard during the hottest days of summer. This form requires two years to come to its full maturity, but its cousin, the seventeen-year locust, requires, typically, seventeen years from the time the eggs are laid until the animals are ready to lay another series of eggs. These eggs are laid in the twigs of trees. The young when hatched from these eggs drop to the ground, and, burrowing beneath its surface, spend the next seventeen years* sucking the juices of the roots of the trees.

Another group of Homopterans are the ''spittle-insects,'' small forms which, settling upon a blade of grass or twig of shrub, soon surround themselves with a frothy mass. They suck the juices of the plant, and after having taken out what they desire eject the rest as a mass of foam. Examine one of these bits of froth and you will find the immature bug inside. Allied to them are the tree-hoppers and leaf-hoppers, so common and so injurious to vegetation.

The plant-lice deserve a little more attention. They occur on almost every kind of plant, sucking its juices and reproducing as rapidly as possible. One does but little damage, but the havoc wrought by thousands is very considerable. In the summer the colonies of these forms will be found to be largely wingless, and these wingless forms are all females capable of reproduction without males. In some species they lay eggs, in others they bring forth living young. These in time reproduce in the same way, and so rapidly do they increase that one plant-louse may be the progenitor of 100,000,000 in five generations. At the close of the season appear the true sexual forms, the males always winged. These sexual forms produce eggs which last through the winter. All of the plant-lice are destructive to vegetation, and some, like the Phylloxera of the grape, are extremely so.

* In the South the period is thirteen years, in the North seventeen.

The scale-bugs or bark-lice are also serious pests, doing great damage to fruit-trees, etc. The males are winged, but the female is scale-like and adheres closely to the branch or the fruit, sucking its juices. Oranges and lemons are frequently covered with these forms. A few, however, are of value to man. The pigment carmine is made from the dried bodies (cochineal) of a scale-louse of the cactus, while lac—from which shellac is prepared—is the secretion of a tropical tree-inhabiting species.

Besides the Heteroptera and the Homoptera, the Hemiptera embraces a third division, the PARASITA, or lice. These are all wingless forms, living as parasites in the hair and on the skin of mammals, and sucking the blood of their hosts.

LEPIDOPTERA (MOTHS AND BUTTERFLIES).

The millers, moths, and butterflies are grouped together as Lepidoptera, and all agree in having four membranous wings covered with dust-like scales, in having a long sucking "tongue" formed of the two maxillæ, and in having a complete metamorphosis (p. 188) in which there hatches

FIG. 84.—Army-worm, larva of *Leucania unipuncta*, showing five (pairs of) abdominal legs.

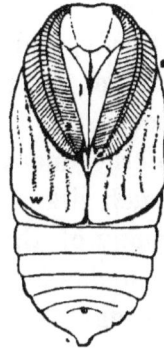

FIG. 85.—Pupa of a Bombycid moth *a*, antenna; *l*, first pair of legs ; *w*, wings.

from the egg a worm-like larva. This stage is commonly known as a caterpillar or "worm," but it differs from all true worms in having legs, and those who wish to call things by their true names should never speak of them as worms. These larvæ always have sharp jaws and simple eyes, and are provided with from eight to sixteen legs. Of these, three pairs are on the thoracic segments, while the abdomen has from one to five pairs. These larvæ, when they hatch from the egg are small, but by feeding they grow, increase in size being rendered possible by frequent moltings of the skin. At last there comes a molt by which the appearance is greatly changed and the pupal stage is reached. In the **pupa** the abdominal legs are lost, the body is shortened and covered with a harder skin, in which one can trace the legs, antennæ, and wings of the future moth or butterfly, folded over the breast. Many caterpillars of the moths, as a preparation for **pupation**, spin silken nests or cocoons, the silk being the product of glands which empty into the mouth. The pupæ of butterflies have usually no such silken protection, but are free. From the fact that many of these butterfly pupæ are marked with patches and spots of gold, they are frequently called **chrysalides** (sing. **chrysalis**).

The pupal stage lasts for some time (months), during which no food is taken and no motion possible except of the abdominal rings; then the pupal skin is molted and the perfect insect (**imago**) emerges. In those forms which have a cocoon the silken threads are softened by fluids secreted by the imago, and in some there are hooks at the bases of the wings which aid in tearing an opening for the escape of the moth.

When the imago first comes out it is soft and flabby, and the wings are soft bags. They are rapidly distended by blood pumped into them, and, held expanded, are rapidly dried by the air into efficient organs of flight. The wings are covered with scales, and to these is the color-pattern due. These scales are merely modified hairs like those which cover the whole body. When removed the wing is seen to have a framework of supporting "veins" which are really not veins at all. These veins vary greatly in their arrangement in different moths and butterflies, and are used as a basis of classification.

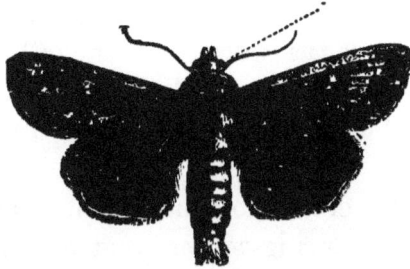

FIG. 86.—Army-worm moth (*Leucania unipuncta*). From Riley.

While the larvæ are biting insects, the adult is adapted for taking liquid nourishment by means of a so-called "tongue" which when not in use is coiled beneath the head like a watch-spring. This tubular structure, which, in function, is so like the beak of the bugs, is much different in structure, as it is formed by the union of the two maxillæ, while the other parts—labrum, mandibles, maxillary palpi, and labium, are present, but in a more or less reduced condition.

There are two great divisions of the Lepidoptera, the butterflies and the moths of common language. The day-flying butterflies hold the wings erect over the back when at rest, and they have the antennæ enlarged into clubs at the tip. In the moths, which are mostly nocturnal, the wings are carried nearly horizontally when at rest, and the antennæ, while frequently feathered, are never clubbed.

Among the smallest, and at the same time the most troublesome, of the moths are those pests, the clothes-moths and their relatives, which do such damage to woolen goods, furs, etc. These are among the few larvæ of moths which have left a vegetarian diet and taken to food of animal origin. Another exception is found in the bee-moth, the larva of which is found in apiaries, feeding upon the wax and spinning its silk all through the comb.

Of the leaf-rolling moths the codling-moth is the best known. Its larva is the worm so frequently found near the core of apples. Other allied species tie the leaves of apple-trees, roses-bushes, etc., together and live in the nest thus formed.

The Geometrids include those moths whose larvæ are commonly known as measuring-worms from their looping gait. All of these are pests, and the canker-worms exceed all the rest in this respect. These are especially noticeable from the fact that the adult female is wingless.

The sphinx-moths or hawk-moths are large narrow-winged forms, the larvæ of which are injurious to many plants. From the attitude assumed by some larvæ when at rest the name sphinx was applied to the group ; the other common name, hawk-moths, has reference to their powers of flight.

FIG. 87.—Sphinx-moth (*Everyx myron*). From Riley.

Another group of moths are known as Bombycids. While some of these are unmitigated pests, others are of value to man, the silkworms leading in this respect. These are, in fact, the most valuable of all insects. The true silkworm is a native of China, but has been distributed to all of the warm parts of the earth. Like other caterpillars, they form their cocoons, and then these are heated to kill the pupa and the silk of the cocoon is unwound, and after proper treatment becomes the silk of commerce. We have several species of silkworms in this country some of which make a stronger silk than the Chinese species; but although a few articles have been made from it, it has no economic importance. These large American silkworm-moths are known as Polyphemus-, Promethea-, Cecropia- and Io moths, and they, together with the beautiful green luna-moth, are great favorites with collectors.

The skippers are a group of small butterflies in which the clubbed antennæ are bent into a hook at the tip. They are called skippers on account of their jerky flight.

The swallowtails are well-known forms of butterflies in which the hind wings are prolonged into tails, whence the name. The larvæ of these forms are usually brightly colored, but they are protected by a pair of "stink-horns," which they can project at will from the region of the neck, and which give off, in most cases, a most offensive odor.

Another group of butterflies, whitish-yellow or orange in color, are typified by the cabbage-butterflies. We had some of these which were bad enough; but a few years ago the European cabbage-butterfly came to this country and became the greatest pest of all our butterflies.

Of smaller size—the most delicate of all our butterflies—are those forms which have received the common names of the blues, the coppers, and the hair-streaks, from their predominant colors and from the ornamentation of the wings.

Of larger size are the group of "four-legged" butterflies, so called because the first pair of legs are so small as to be of no use to the animal.   Of these forms there are hundreds of species, including the milkweed-butterflies, the

FIG. 88.—A four-legged butterfly (*Argynnis aphrodite*), under side shown on right.

painted-beauty, the mourning-cloak (the first butterfly to appear in the spring), and numbers of others, the catalogue

FIG. 89.—White Mountain butterfly (*Œneis semidea*).

of the names of which would prove dry reading.   Only one needs more mention.   This is the White Mountain butterfly, found only on the tops of the White Mountains, on the tops of the higher peaks of Colorado, and in Labrador.   It is supposed that this form is a remnant of an Arctic fauna which extended over the Northern United States when the country was covered by the great ice-sheet (see Geology), and on the retreat of the glacier these colonies were left stranded upon these points as the only places cold enough for them.

COMPARISONS.

With two columns, one for grasshopper, squash-bug, beetle, and butterfly, and the other for crayfish and sow-bug, answer the following questions :*

1. How many pairs of antennæ ?
2. How does the animal breathe ?
3. How many segments in the head-region ?
4. How many walking-feet ?
5. Are there appendages on the abdomen ?
6. Where are the reproductive openings ?
7. Are any appendages two-branched ?

* If the points have not been made out for all forms, answer for those about which you know.

CRUSTACEA.

The crayfish and sow-bug may be taken as types of the Crustacea, or crab-like forms. These all have two pairs of antennæ or appendages in front of the mouth; they have a varying number of segments at the front of the body, covered by a common shell or **carapax**, and, excepting gill-less microscopic forms, they all breathe by means of gills attached to some of the feet.

The number of segments in the body varies; in the higher groups it is constantly twenty, but in the lower it may fall far short of, or far exceed, that number. The regions also vary in extent and cannot be compared throughout the group. Taking the segments connected with the senses and with eating as constituting the head, this region may contain as few as five or as many as eight segments. Not infrequently the head and the next region of the body are united so that they are called a cephalothorax. The abdomen is usually well developed, but it may be reduced to a mere stump, as in the barnacles. Any of the segments except the last one may bear appendages. Those most usually present are two pairs of antennæ,* a pair of mandibles, two pairs of maxillæ, and a varying number of maxillipeds and walking-feet.

If we study these appendages in the young, or in the adult of some forms, we find that they each consist of a

---

* One pair is very small in the sow-bug, but it can be seen with a lens.

203

basal joint, bearing two jointed branches, the exopodite and endopodite. With growth of the animal the exopodite frequently disappears.

The gills by which most Crustacea breathe are thin outgrowths of the body, usually closely connected with some of the appendages, either of the thorax or of the abdomen. In shape they may be plates or plumes or sacs, but all are traversed by blood-vessels so that the blood is brought in close proximity to the water. In some cases these gills hang freely into the water, in others they are placed in special gill-chambers, and then there is an arrangement of parts for pumping fresh water over them. In the terrestrial Crustacea these gills still serve as breathing-organs, as in the sow-bugs, and are constantly kept moist. In some of the lower Crustacea there are no special organs of respiration, the thin walls of the body affording sufficient surface for the purpose.

The alimentary canal is nearly straight, and there is usually a chewing-stomach in which the food is ground by hard teeth in the walls, and beyond this there is frequently a straining-stomach. A large so-called liver is always present, pouring digestive juices into the alimentary canal behind the stomach. The eyes are either simple or compound. In the simple eyes there is a single lens for the whole structure, while the compound eyes are composed of many separate eyes, each with its own lens. In some cases these eyes are placed on jointed stalks, at others they are in the walls of the head. Ears have been found in some forms. Usually they are sacs in the base of the antennulæ, but in the opossum-shrimps they occur near the end of the abdomen. The hairs which occur over the body are organs of touch, and possibly some of them serve as organs of taste and smell as well.

A heart is lacking in a few forms. When present it is dorsal in position, but may be either in thorax or abdomen. It may be a long tube with several chambers, or a short thick muscular organ without divisions. The blood returning from the gills enters the heart and is forced thence to all parts of the body, a condition quite different from what we found in the fish. It does not flow throughout its course in closed vessels, but escapes from them and comes into large spaces (**lacunæ**) between the various organs and muscles, and from the largest of these lacunæ, near the floor of the body, it again goes to the gills.

In the Crustacea there are excretory organs (**nephridia**) which open to the exterior entirely independently of the alimentary canal. In the higher Crustacea (crayfish, etc.) these nephridia are known as "green-glands" and open at the base of the antennæ (second segment); in the lower Crustacea they are called "**shell-glands**" and open at the base of the second maxillæ (fifth segment).

The sexes are separate in all except the barnacles, and the ducts of the reproductive organs open to the exterior in the thoracic region, never in the abdomen. In almost all forms the eggs are carried about by the mother until they are hatched. In almost all the lower Crustacea the young escapes from the egg in a very immature condition, known as a **Nauplius**, a name given years ago under the belief that it was an adult. The nauplius has an unsegmented body, a single median eye, and only three pairs of appendages—antennulæ, antennæ, and mandibles—the antennulæ being solely sensory, while antennæ and mandibles are used in both swimming and eating. In the higher Crustacea the nauplius stage is passed in the egg, and the young hatches in a more advanced condition—sometimes closely like the adult in all except size. Growth is allowed for by frequent molts of the external cuticle of the body.

Over 10,000 species of Crustacea are known, almost all of them aquatic, and the majority marine. Only a few, like the sow-bugs and land-crabs, live on the land. A few are vegetarians, some are parasites on other animals, but the majority are scavengers, feeding on decaying organic matter. The Crustacea may be conveniently divided into two "subclasses": Malacostraca and Entomostraca.

SUBCLASS I.—MALACOSTRACA.

This group contains the larger and higher Crustacea, in which the body consists of twenty somites,* all of which except the last (telson) may bear appendages. Compound eyes are usually present; and the nauplius stage (p. 208) is usually passed in the egg. Besides several unimportant groups, this subclass contains the orders Decapoda and Tetradecapoda.

ORDER I.—DECAPODA (p. 164).

ORDER II.—TETRADECAPODA (p. 168).

SUBCLASS II.—ENTOMOSTRACA.

This division contains a large number of forms, mostly small, or even microscopic in size. The number of body-segments is usually less than twenty, but occasionally there may be many more. Some are decidedly shrimp-like in form, but in others parasitic habits have resulted in such changes that there is little external resemblance to a crayfish or a crab. In fact, this degeneration may go so far in

---

* Twenty-one in Nebalia.

certain fish-parasites (so-called fish-lice) that the adults
would never be suspected of being crustaceans were it not

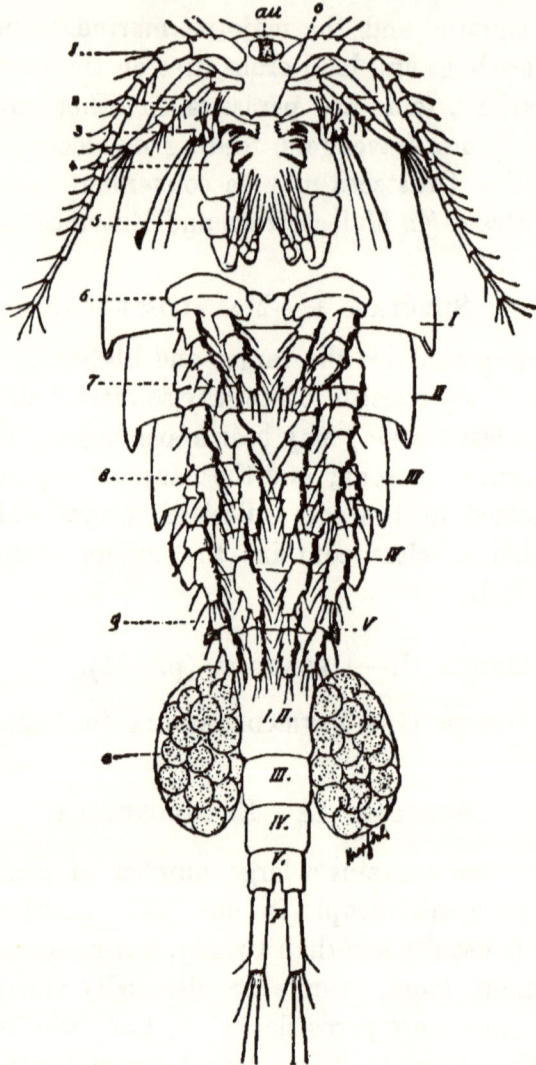

FIG. 92.—An Entomostracan (*Cyclops*). From Hertwig.

for the young. When the development is studied, these,

like all other Entomostraca, are found to have a free-swimming nauplius stage.

The only ones of the Entomostraca (aside from the fish-parasites) which have received a common name are the barnacles, so familiar at the seashore. In these the body is enclosed in a hard calcareous shell, which is either directly attached to some solid support, as in the acorn-barnacles, or there is a fleshy support, as in the goose-barnacles. Inside the

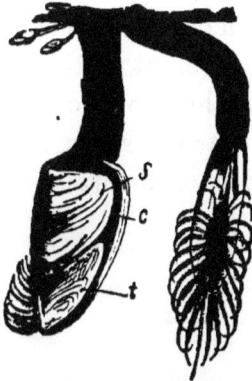

FIG. 93.—Goose – barnacles (*Lepas anatifera*). After Schmarda.

shell is the animal, and a cursory examination of its two-branched feet and its other features would con-vince any one that these forms are truly crustacean.

Mention should be made here of a large group of ex-tinct animals, the Trilobites, which recent investigations have shown to be crusta-ceans, but which cannot be more definitely placed within that group. They agree with

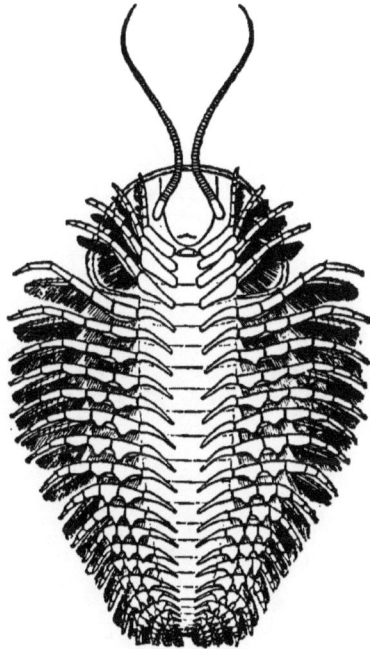

FIG. 94.—Restoration of the under surface of a Trilobite, showing the appendages. After Beecher.

neither Entomostraca nor Malacostraca in their structure. They have a flattened body, in which head, thorax, and abdomen are readily distinguished, and in which both thorax and abdomen consist of an axial portion, and two lateral regions or lobes, whence the name of the group. The head bears a pair of compound eyes, a single pair of antennæ, and four pairs of appendages, which served at once for walking and for taking food. Each segment of thorax and abdomen supports a pair of two-branched appendages. Trilobites appear in the earliest fossil-bearing rocks, and the group died out soon after the period of coal-formation (in the Permian).

HEXAPODA (Insects).

The group of Insects contains more species than all the rest of the animal kingdom together, a conservative estimate placing the number of distinct forms at over half a million. Yet all of these agree in certain essential points. Thus, in all, the body is divided into three regions, head, thorax, and abdomen, and of these the thorax alone bears organs of locomotion. Three pairs of legs are always present (whence the name Hexapoda—six-footed—given to the group). Of wings there may be one or two pairs. The head bears four pairs of appendages, one pair (the antennæ) being sensory ; the others (mouth-parts) being used in eating. Breathing is by means of tubes (**tracheæ**) which open on the sides of the body and which penetrate to all parts of the interior. The sexes are always separate, and the reproductive organs open at the hinder end of the body just beneath the vent.

In the head no evidence of segments is seen, except as shown by the appendages. The antennæ, of which there are only a single pair, are sensory in function. In many cases they clearly bear organs of smell, and in some they may also be hearing-organs. In the primitive condition the mouth-parts are fitted for biting and eating hard substances, the mandibles being strong jaws, while the maxillæ and labium serve to hold the food in place. These latter bear jointed prolongations—the palpi—which are sensory. In other insects these mouth-parts are modified and united

213

into a sucking-tube which frequently is a piercing-organ of no mean capabilities.

The **thorax** is composed of three segments, named, from in front backwards, the **prothorax, mesothorax,** and **meta-thorax.** Of these the first is frequently movable. Each segment bears a pair of legs, made up of several joints, the number varying according to the number in the "foot" (tarsus), the rest of the member usually consisting of four joints. On the dorsal surface of the meso- and metathorax occur the wings, the characters of which are largely used in the classification of insects. They are entirely lacking in the lowest insects (Thysanures) as well as in individuals of other groups, as ants, many parasites, and the females of certain moths. In the flies the posterior wings are greatly reduced, so that they appear like a pair of knobbed hairs, termed "balancers," since if they be removed the fly cannot control its motions. Frequently both pairs of wings are used in flight, but in certain groups the front pair are much thickened and hardened, so that they are converted into wing-covers (**elytra**) which protect the hinder wings when at rest.

The abdomen is normally composed of ten segments, but this number may be reduced. In some insects the abdomen joins the thorax by its whole width, while in others it is contracted in front to a slender stalk as in the wasps. The appendages of the abdomen, in the adult, are never locomotor in function. In the lowest insects rudimentary appendages may occur on all segments of the abdomen, but in the higher groups only three pairs, at most, occur, and two of these are modified into an organ (ovipositor) for laying the eggs. In the bees, wasps, etc., the ovipositor is at the same time an offensive weapon, the sting.

The alimentary canal has few convolutions. Into the

mouth-cavity open the salivary glands. In those forms which eat solid food a "chewing-stomach" with hard horny teeth occurs. Behind this comes the true stomach, and following this the intestine, to which are attached a varying number of **Malpighian tubes** (2–100 or more) which, like the kidneys of higher forms, serve to carry away nitrogenous waste from the body.

The circulatory organs are poorly developed. A dorsal tube, or heart, is present, lying above the alimentary canal, and this pumps the blood forward, into an **aorta** of varying length. Soon, however, the blood leaves this tube and flows between the muscles and viscera and finds its way to the hinder part of the body, where it again enters the heart through openings in its sides. This imperfection in the blood-vessels is compensated for by the peculiar character of the organs of breathing (respiration). These consist of a number of tubes (**tracheæ**) which open to the outside by paired openings (**spiracles**) in the sides of the body. These spiracles occur in the thorax and abdomen, and never exceed a pair to a somite, and from three to ten pairs may occur. Internally the tracheæ branch again and again, until the finest twigs penetrate to every part of the body. Frequently the various tracheæ are connected on either side of the body, and in the strong-fliers these connecting tubes are enlarged into air-sacs, which thus render the body lighter. Air is drawn into the tracheæ by the enlargement of the abdomen, and thus reaches all the of tissues of the body. Since breathing is accomplished through the spiracles in the sides of the body, one can see that one cannot readily kill an insect by putting chloroform on its head.

The nervous system consists of an enlargement or "brain" in the head, in front of the mouth, and from this nerves go to the eyes and antennæ, while a stronger nerve-cord passes on either side of the gullet, to unite in a second enlargement (ganglion) behind. Thus, as will readily be understood, the alimentary canal passes through the nervous system, a condition which is totally different from anything found in the vertebrates. Behind the infra-œsophageal ganglion a double nerve-cord extends along the floor of the body, connecting a series of similar ganglia. In the lower insects there is a ganglion in each segment, but in the higher these tend to move forward and to unite with each other into a few masses or compound ganglia.

The eyes are always on the head. In the adult insects compound eyes are usually present, and besides these there may also be simple eyes. In the latter there is but a single lens, while the compound eyes are composed of many distinct visual structures, each with its own lens. Organs, which are regarded as ears, occur in various forms. In the grasshoppers these organs are on the base of the abdomen; in the crickets, on the legs; in many groups the antennæ are supposed to have auditory powers. Taste resides chiefly in the lower lip, while touch, though found all over the body, is especially developed in the antennæ and the palpi of labium and maxillæ. In some insects the sense of smell is strongly developed, and there is reason to believe that the olfactory organs are in the antennæ.

The group of Insecta may be subdivided in two ways, accordingly as different characters are employed. If we follow one method the mouth-parts form the basis of division, and we have a *mandibulate* group in which the jaws are fitted for biting, as in the grasshopper and beetle; while in the *haustellate* group the mouth-parts are no longer fitted for biting, but form a tube through which liquid food is sucked, as in the bugs and butterflies.

The second method of subdivision depends upon the

facts of life-history. In the first or *ametabolous* group the young leaves the egg with much the general shape of the adult, and the growth is gradual, without any sharply marked lines between the different stages. Such is the case with the grasshopper and the bug. In the other or *metabolous* group we can distinguish three stages sharply marked off

FIG. 95.—Colorado potato-beetle (*Doryphora decemlineata*). *a*, eggs; *b*, larva; *c*, pupa; *d*, adult.

from each other—larva, pupa, and adult. These are exemplified in the beetle and butterfly.

These two classifications do not agree, as can be seen from the following tables:

MANDIBULATÆ.	HAUSTELLATÆ.
Thysanura.	Hymenoptera.*
Orthoptera.	Hemiptera.
Pseudoneuroptera.	Lepidoptera.
Neuroptera.	Diptera.
Coleoptera.	

* The Hymenoptera have the mouth-parts adapted for both biting and sucking.

AMETABOLA.	METABOLA.
Thysanura.	Coleoptera.
Orthoptera.	Neuroptera.
Pseudoneuroptera.	Hymenoptera.
Hemiptera.	Lepidoptera.
	Diptera.

As will be seen from the foregoing tables, the group of Hexapoda, or Insecta, is subdivided into nine groups or orders.*

ORDER I.—THYSANURA.

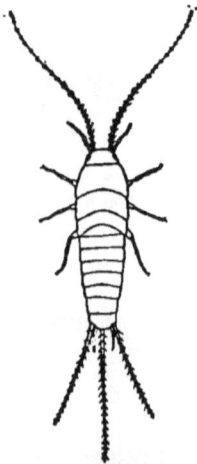

These are small wingless insects without any general common name except those of "bristle-tails" and springtails, which have been manufactured for them. The springtails live in damp places—in cellars, under leaves in the forest, etc., and they have a spring beneath the body by means of which they can jump to great distances. The bristletails have the body terminating in two long filaments. To this last group belong some pests known commonly as "silverfish"—soft-bodied shining forms, which eat paper, starched clothing, etc. Aside from this silver-fish or "fish-moth" the group has little general interest; but to the naturalist it is very interesting because it is so primitive.

FIG. 96.—"Silver-fish" (*Leptisma saccharina*).

ORDER II.—ORTHOPTERA (see p. 181).

* Many authorities recognize more orders than these, the difference chiefly lying in the extent to which the Neuroptera and Pseudoneuroptera are subdivided.

ORDER III.—PSEUDONEUROPTERA.

These forms, like the Orthoptera, have biting mouth-parts, and have a gradual change from the young to the adult, but they differ from those forms in having both pairs of wings alike, usually very thin and transparent, with very numerous veins, and not capable of being folded like those of the Orthoptera. There are two divisions of these Pseudoneuroptera. In the first the younger stages are passed in the water, in the second on land.

Examples of the first are seen in the dragon-flies (ODONATA); their larvæ live in the water, where they feed upon other insects, etc., and especially on the larvæ of mosquitoes. When the adult stage is reached and they take to the air, they are veritable dragons, feeding upon insects, which they catch on the wing. Here, too, belong the May-flies or day-flies with an aquatic life of from one to three years, a life in the air of but a few days, or even a few hours. These May-flies often appear in great numbers in the cities near the Great Lakes.

The celebrated white ants or termites may represent the forms with a solely terrestrial life-history. These are not "ants" at all in the true sense of the word, but they resemble them in several points. They form large colonies consisting of several distinct "castes" with different structure. Only the kings and queens are winged, and only these are capable of reproduction. Besides these there are "workers" and "soldiers." The workers build the nests, gather the food for the whole colony, and bring up the young. The soldiers have enormous heads, and protect the others. The termites are miners, and make their burrows beneath the earth and inside of dead wood. They avoid the light, and where they cannot otherwise make their way they build covered ways, sometimes for hundreds of feet. They feed upon dead wood, and will sometimes utterly eat out the inside of the timbers of a house, leaving posts and joists but a mere shell. They are comparatively rare in colder climates, but in the tropics

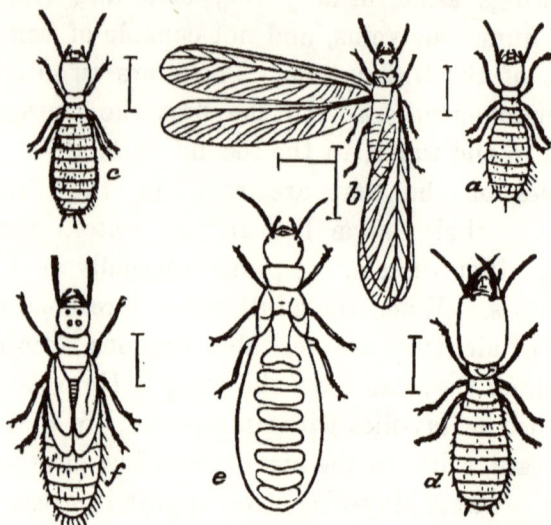

Fig. 97.—White ant (*Termes flavipes*). *a*, larva; *b*, winged male; *c*, worker; *d*, soldier; *e*, queen; *f*, pupa. From Riley.

they become a terrible pest. The queen is kept a prisoner in the nest, is fed by the workers, and develops so many eggs that her abdomen is swollen out of all proportion. As the eggs escape they are cared for by the workers.

ORDER IV.—NEUROPTERA.

These forms have the wings much as in the Pseudoneuroptera, the mouth-parts for biting or much reduced, but they have a complete metamorphosis. The majority of these forms are inconspicuous, and their existence is hardly recognized except by naturalists. Here belong the "dob-

sons," or hellgrammites, larvæ of a large insect, which are
used as bait by anglers.   Here, too, belong the ant-lions,

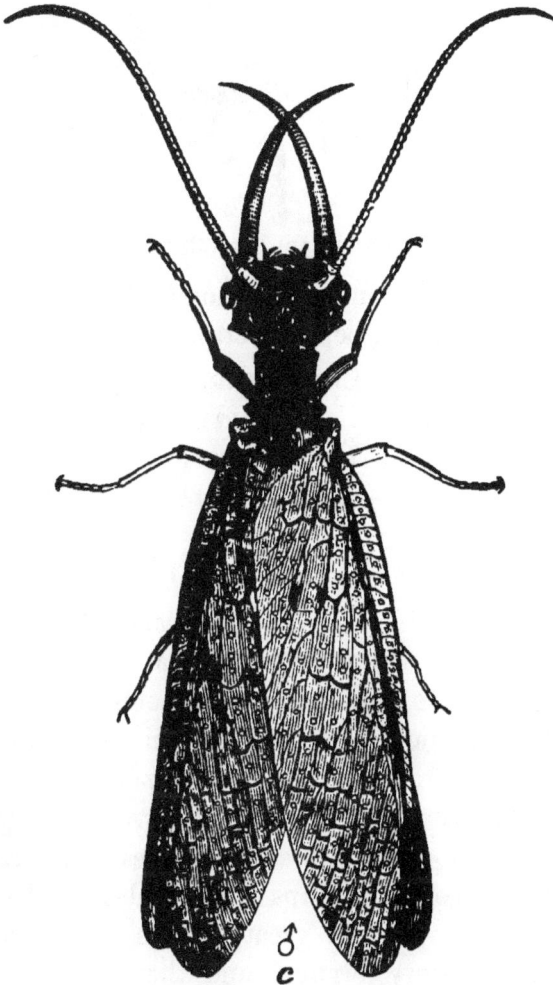

FIG. 98.—Adult male hellgrammite (*Corydalis cornutus*).   From Riley.

which build little pitfalls for the ants on which they feed.
Last to be mentioned are the caddis-flies, the aquatic larvæ
of which protect themselves by building cases of stones,
sticks, etc., in which they hide and which they carry about

with them in their search for food. These caddis-flies, in the adult stage, have the mouth-parts much reduced, and

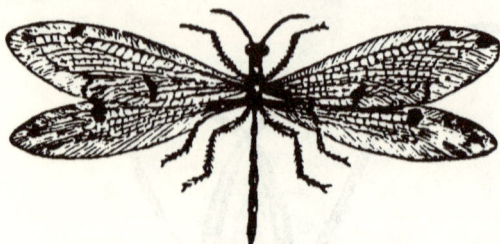

FIG. 99.—Adult ant-lion (*Myrmeleon*).

are supposed to represent pretty closely the ancestors of the butterflies and moths (Lepidoptera).

ORDER   V.—COLEOPTERA (see p. 184).
ORDER   VI.—HYMENOPTERA (see p. 188).
ORDER VII.—HEMIPTERA (see p. 194).
ORDER VIII.—LEPIDOPTERA (see p. 199).

ORDER IX.—DIPTERA (Flies).

This order contains the true flies, and these forms are sharply marked off from other insects. The name means two-wings, and the flies have but a single pair of these organs, while on the metathorax is a pair of knobbed hairs, the so-called balancers. The mouth-parts are fitted for sucking. The larvæ, commonly known as maggots, are worm-like, lack feet, and in some species even lack a distinct head. In some the pupa is motionless, but in others, as in the mosquito, it has great powers of motion. The balancers are sensory organs, and they also serve as a means of maintaining the equilibrium, for if they be cut off from a fly, the animal can no longer direct its motions.

The group of flies is very large in number of species, some being beneficial, while others are decided pests. Among the latter are those forms which feed upon other insects, as well as those which in their larval stages feed upon decaying organic matter.

Most familiar of all is the common house-fly. This lays its eggs in horse-manure, each female producing about 150 eggs. In about ten to fourteen days these eggs become perfect insects, so that with this rapidity of multiplication it is no wonder that flies are abundant towards the end of summer. Allied to this is the blow-fly which lays its eggs in meat and other provisions.

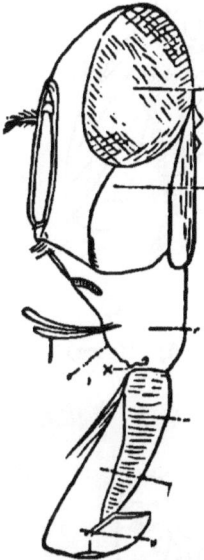

FIG. 100. — Head and Proboscis of blow-fly. After Kraepelin. *e*, egg; *p*, maxillary palpi.

FIG. 101.—Larva (maggot) of house-fly.

The bot-flies are parasitic in various domesticated animals. These flies lay their eggs upon horses, cattle, or sheep, and the larvæ enter the animal and cause serious injury or even death. The horse-bot larvæ are taken into the stomach; the ox-bot or " ox-warble " lives beneath the skin of cattle; and the sheep-bot enters the cavities connected with the nose or even the horns, producing the disease known as "staggers."

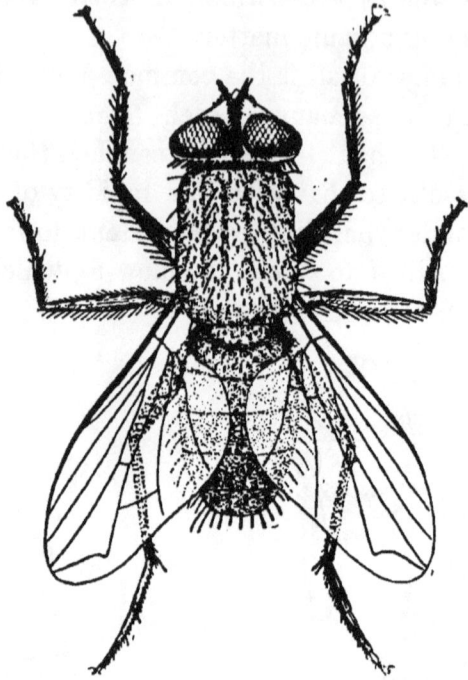

FIG. 102.—Common house-fly (*Musca*).

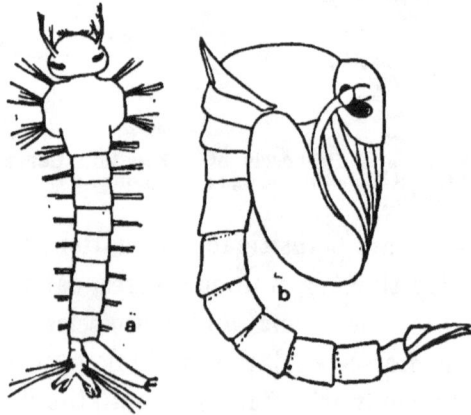

FIG. 103.—Larva (*a*) and pupa (*b*) of mosquito.

More familiar are the mosquitoes, which lay their eggs on stagnant water. The larvæ hatch out and are known as "wrigglers." They pupate beneath the surface, and finally the perfect insect emerges to make itself an unmitigated nuisance about our persons. Many proposals have been made for reducing the number of these torments. The best is, possibly, the pouring of kerosene upon the surface of all stagnant water. This will kill the eggs as they are laid, while it also destroys the perfect insects as they come from the water.

COMPARISONS.

With two columns, one for grasshopper, beetle, squash-bug, and butterfly, the other for crayfish and sow-bug, answer the following questions:

(1) Is the body made up of a series of segments?

(2) Do any of the segments have jointed appendages?

(3) Do you find more than one pair of appendages on one segment?

(4) Are the hard parts (skeleton) external or internal?

(5) Do the jaws work in a vertical or in a lateral plane?

(6) Can the jaws be compared to the other appendages of the body?

(7) Is the heart above or below the alimentary canal?

(8) Is the brain above or below the œsophagus?

(9) Where is the largest part of the nervous system?

(10) Is there any relationship between nerve-enlargements (ganglia) and the external segments of the body?

ARTHROPODA.

The word Arthropoda means "jointed foot," and is very characteristic of all that immense series of forms which, like the grasshopper and the crayfish, have an external skeleton which only permits of motion by a thinning or jointing at intervals. In this way both body and limbs have this jointed appearance, but with the body this jointing or **segmentation** of the external surface is associated with features of internal structure which must have a moment's attention. This external jointing of the body divides it into a series of essentially similar rings or **somites**, and in each of these we find parts of all the internal organs. That is, the segmentation is not confined to the external surface, but is characteristic of all parts.

In an ideal arthropod each of these segments would be an exact repetition of its fellows, but in nature we find that certain segments or parts of certain segments become over-developed (**hypertrophied**), and this produces an under-development (tendency towards **atrophy**) in others. Thus every segment in our ideal arthropod would bear a pair of jointed appendages, but our studies have shown us that these appendages are frequently atrophied on some of the segments. Again, there is a tendency in some regions, and especially in the head, for a more or less complete fusion of segments, so that the number can only be ascertained by the appendages or by the features presented in development.

Usually these segments can be grouped in regions, of which, at most, three can be distinguished: in front the head; next, the thorax; and behind, the abdomen. The head is largely concerned in the taking of food, and is the seat of the special senses. The thorax is the locomotor

Fig. 104.—Diagram of grasshopper showing the body divided into the three regions: head, thorax, and abdomen.

region, while in the abdomen the primitive segmentation is most marked.

Through the body as an axis runs the alimentary canal, the mouth being on the under surface of the head, while the vent is at the tip of the abdomen. Above the digestive tract lies the heart, which in some forms has a chamber in each of several somites of the body; that is, the heart is segmented. On the floor of the body, below the alimentary canal, is the nervous system, which exhibits this segmentation in a more marked degree. In each segment there is a paired enlargement or **ganglion** from which nerves go to the various organs of the segment. These ganglia of the successive segments are connected with each other by a double nerve-cord, so that all are in communication with each other. At the front end of the body one of these nerve-cords passes on one side of the œsophagus, the other on the other, and above it they unite with a large compound ganglion, the so-called brain. In this way a part of the nervous system is brought above the alimentary canal, while the rest lies below. In other words, the digestive tract passes through the nervous system, a condition which is without parallel in the vertebrates, but which is usually met with in the non-vertebrate animals.

The organs of respiration are never connected with the alimentary canal. They are always developments of the surface of the body. In the case of gills we have outgrowths of the body-wall, usually much folded or divided to afford additional surface, and in these are blood-vessels. In the case of gills, then, we may say that the blood is brought to the oxygen dissolved in the water for that exchange of gases (carbon dioxide and oxygen) upon which respiration depends. With tracheæ, on the other hand, the respiratory surface is obtained by a forcing of the external surface into the deeper parts, much as one might invert the finger of a glove into the palmar region. In the tube thus formed air can enter, and thus the oxygen is brought to the blood and other tissues of the body.

The Arthropoda are by far the largest group of animals, the number of forms living to-day being estimated from half a million to a million or more.

The Arthropoda are subdivided into three groups or "classes": Crustacea, Acerata, and Insecta.

Class I.—CRUSTACEA (p. 206).

Class II.—ACERATA.

In these arthropods the body is divided into two regions, a cephalothorax in front and an abdomen behind. The cephalothorax bears the eyes (of which there may be several pairs) and six pairs of appendages, none of which can be considered as antennæ. The abdomen may have or may be without apparent appendages. The respiratory organs are confined to the abdomen, and in their development are always connected with the abdominal limbs. They may be of three kinds: (1) External gills borne on the abdominal legs; (2) internal sacs (lungs) with numerous leaf-like folds; (3) air-tubes or tracheæ, strikingly like those of the Insecta, but with a different history. The reproductive organs open near the middle of the body.

Subclass I.—Merostomata.

Here belong the horseshoe crabs of our east coast (and a number of fossil forms), which breathe by means of leaf-like gills, which have both simple and compound eyes, and which have the bases of the walking-feet of the cephalo-thorax modified to serve as jaws. Recent investigations show that the horseshoe crabs are not related to the true crabs, but are to be rather closely associated with the scorpions. These forms live in the sea, feeding on worms, etc., found in the sea-bottom, coming to the shore in spring and early summer to lay their eggs. The horse-shoe crabs are without any economic importance, as they are useless as food, but they are extremely interesting to the naturalist, as they are the last remnants of forms which were once abundant in the seas of past times.

Subclass II.—Arachnida.

With few exceptions, the Arachnids are terrestrial forms. They breathe by internal lungs or by tracheæ, and they lack compound eyes. There are several orders of Arach-nids, but only a few of them need be mentioned here, as some are inconspicuous, while others occur only in the warmer regions of the globe.

Order I.—Scorpionida.

The scorpions have a single pair of jaws (mandibles) and a pair of large pincers, much like those of lobster or crab. The long abdomen is distinctly jointed, the seven basal joints being much larger than the terminal five.   The abdomen ends in a very efficient poison-sting.   On the lower surface of the basal abdominal segments are the openings to four pairs of lungs.   Scorpions are not found in cold

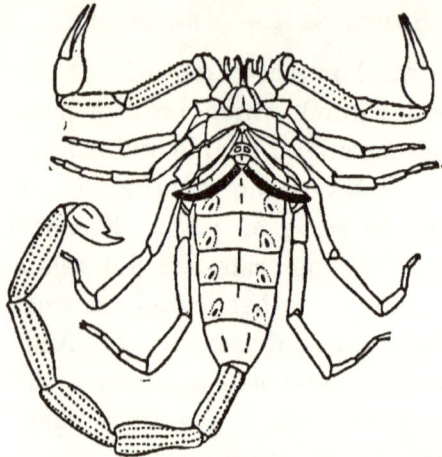

FIG. 105.—Under surface of scorpion (*Buthus*) showing the combs and outlines of the lung-sacs.

climates, but in the warmer regions they abound, and their stings, which rarely prove fatal to man, renders them unpleasant companions.

ORDER II.—ARANEIDA.

The Araneida, or spiders, have the cephalothorax and abdomen unsegmented, but sharply separated from each other by a narrow waist.   In front are the poison-jaws (mandibles), each with a poison-gland inside.   At the tip of the lower surface of the abdomen are two or three pairs of **spinnerets.**   These are modified appendages with numbers of small openings at the tip.   Connected with each spinneret is a gland which secretes a fluid with the property of hardening as soon as it comes in contact with the air.   This is

forced out at will through the spinnerets, and forms the
silk with which the spiders wind their prey, wrap up their
eggs, and build those marvellous webs, interesting to all

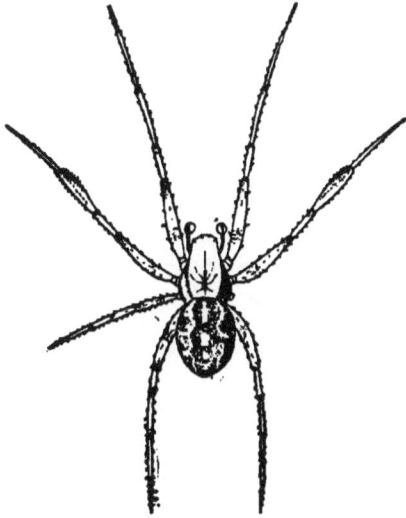

Fig. 106.—Round-web spider (*Epeira insularis*). After Emerton.

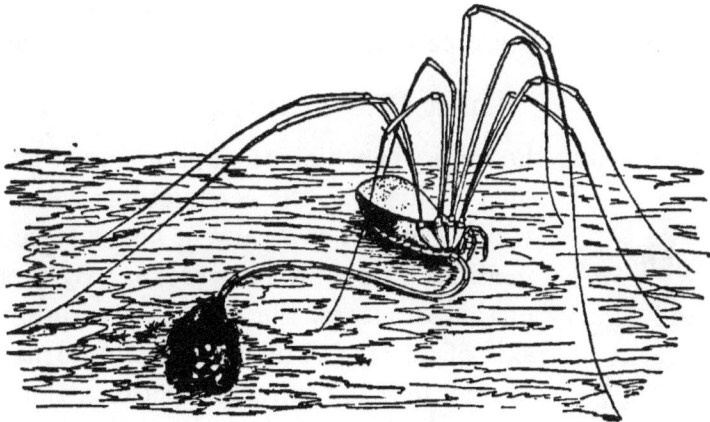

Fig. 107.—A harvestman laying eggs. After Henking.

except the housekeeper. The poison-jaws are strong, and
venomous enough to kill the insects upon which these ani-

mals feed; but the alleged cases of serious or fatal poison-
ing of man as the result of spider-bites need authentication.

Order III.—Phalangida.

This name is given to the animals familiarly known as
"harvestmen" and "daddy-longlegs," with small bodies
in which there is no waist between thorax and abdomen,

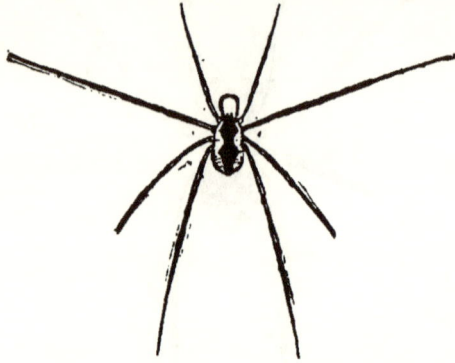

Fig. 108.—Harvestman (*Phalangium pictum*).

and with extremely long legs. These forms feed upon
small insects, but are perfectly harmless to larger animals.

Order IV.—Acarina.

Here belong the mites, in which the unsegmented abdomen is fused to the cephalothorax, and in
which the first two pairs of appendages are
modified into a piercing-organ. By means
of this structure, the ticks burrow into the
skin of cattle or of man, the itch-mite makes
its way into the thin skin between the fingers, and the red mite sucks the juices of
plants. As a rule the Acarina are parasites, and hence the group is largely made
up of pests.

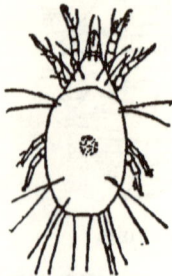

Fig. 109.—Cheese-
mite, enlarged.

Class III.—INSECTA.

In the Insects there is a distinct head consisting of four segments; respiration is effected by tracheæ opening along the sides of the body, while the reproductive organs open near the tip of the abdomen.

Subclass I.—Hexapoda (p. 213).

Subclass II.—Chilopoda (Centipedes).

In the Chilopods, which include the centipedes and similar forms, the head is succeeded by a long series of body-segments, each with a pair of locomotor appendages (legs), and with no distinction between thorax and abdomen. Most of the group are carnivorous, and the larger forms, at least, are provided with poison-glands which open in the last pair of cephalic appendages. The chilopods of northern latitudes are small, insect-feeding forms, but in the tropics occur the centipedes, the larger species of which are said to be extremely venomous.

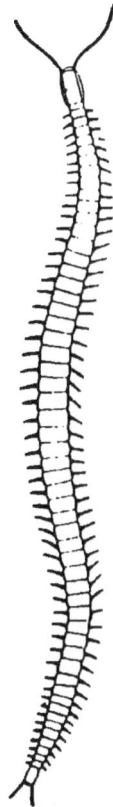

Fig. 110.—A Chilopod (*Geophilus*).

Usually the Chilopods are associated with another group, the Diplopoda (thousand-footed worms), as a class or subclass, Myriapoda, but the differences between them are too great for this. The Diplopods have but three segments in the head, and, after the first three, each segment of the body bears two pairs of legs, while the reproductive organs open far forward. The thousand-legged worms live in moist places, where they feed upon decaying vegetable matter.   They are harmless forms, but several species se-

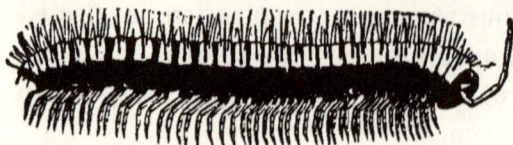

FIG. 111.—A Diplopod (*Spirostrephon*), showing the two legs to a segment. From Packard.

crete a strong-smelling substance, which protects them against their foes.

LABORATORY WORK: EARTHWORM.

The student should be supplied with a live earthworm, and also with a specimen killed by placing in a dish in which is a bit of cloth dampened with chloroform, the whole being covered so as to prevent escape of the fumes. After death the worm should be pinned out straight, and hardened in plenty of alcohol.

Is the body cylindrical throughout? Is it bilaterally symmetrical? Can you distinguish between dorsal and ventral surfaces? Is the body apparently made up of segments? Are they all essentially alike? Draw the worm through the fingers; does it move with equal ease in both directions? Examine the head end for the mouth; is it dorsal or ventral in position? Is the ring (**preoral lobe**) in front of the mouth complete? How is it attached to the next ring? Examine the surface of the body with a lens for bristles (**setæ**). Do you find them on each segment? How are they arranged on the segment? Where is the vent? About one fourth the length of the body from the anterior end notice that certain rings are enlarged and swollen, and that the lines between the segments tend to be obliterated. This is the **clitellum**. How many segments are included in it? The clitellum is a glandular structure to secrete the cases or **cocoons** in which the eggs are laid.

Hold a living worm near the anterior end. Does it project a proboscis from the mouth? Look on the back and see the red dorsal blood-vessel showing through the skin.

235

Study the segments in front of the clitellum, looking for openings of the reproductive organs on the ventral surface. How many pairs of these do you find, and on what segments are they?   Leave a dead worm in water for several hours; can you separate from it an external transparent cuticle?

Draw a worm from the side, being careful to get in the right number of segments, back to the posterior end of the clitellum, and bringing out as many of the points discovered as possible.

Pin a worm, which has been in alcohol, with pins passing through the preoral lobe and the hinder end of the body, in a dissecting-pan. With the scissors open the dorsal wall of the body from just behind the clitellum to the anterior end, taking care to cut through only the dorsal wall. It is best to make this cut just a little to one side of the median line. As you start to lay open the body, notice the partitions (**dissepiments**) running in from the body-wall and holding the parts together. Do these dissepiments correspond in position to the external rings or to the spaces between them? Do they divide up the body into a series of body-cavities? Do the cavities of the right side correspond in position with those of the left?

Cut the dissepiments with the scissors, and pin out the body-wall. This exposes the digestive tract lying in the axis of the body. In it make out the following regions: (1) A pear-shaped enlargement (**pharynx**) occupying about half a dozen segments in front. Notice the muscle-fibres going to the pharynx from the body-wall. (2) A narrower tube (**œsophagus**) leading back through about ten segments from the pharynx, and expanding about segment 16 into (3) a heart-shaped **crop**, which in turn is followed by (4) a second enlargement (**gizzard**) of about the same size. (5)

From the gizzard the **intestine** can be traced back to the vent.

Lying above the alimentary tract is the **dorsal blood-vessel.** From it are given off transverse vessels. Are these in pairs ? Do they correspond to the segments in number and position ? Are any of them enlarged ? In what direction do they go ? Can you find (by tipping the alimentary canal) a **ventral blood-vessel** beneath ? Do any vessels connect with it ?

On the top of the anterior end of the œsophagus are two pear-shaped bodies, the **brain.** Can you find nerve-cords **(commissures)** leading downward and backward from the brain ?

Arising from either side and extending upwards so as to overlap the œsophagus above are lobes of the **reproductive organs.** Draw the parts so far made out, viewed from above, and then cut through the pharynx and carefully lift up the alimentary canal as far back as the beginning of the intestine, cutting it off at that point. Now sketch the reproductive organs, lifting them up to see if any organs occur beneath.

Examine the cut end of the intestine. Is the inside a circular tube ? On the dorsal surface of the intestine see the dark green **chloragogue organ** (a digestive gland, supposed to be something like liver or pancreas in its action).

On the middle line of the floor of the body find the **ventral nerve-cord,** with its numerous enlargements **(ganglia).** Are these latter equal in number to the somites ? Do they occur in or between the somites ? Trace the nervous system forwards, and find out how it connects with the brain. Draw the brain and twenty ganglia of the ventral chain connected together. Just outside the ventral nervous cord find in each segment (except a few anterior)

a minute coiled tube (**nephridium**). These are the kidneys
of the worm, and each opens separately to the exterior
between the rows of setæ.

COMPARISONS.

With columns for Vertebrate, Arthropod, and Earthworm, answer the following questions:

(1) Are paired appendages present?

(2) Do you find an evident body-cavity?

(3) Is the alimentary canal supported by a mesentery?

(4) Is the greater part of the nervous system dorsal or
ventral in position?

(5) Is there any segmentation visible from the outside?

(6) Is there anything which you could call internal segmentation?   If so, what parts are repeated?

(7) Is there an external cuticle?

(8) Does the alimentary canal pass through the nervous
system?

(9) Is there an internal skeleton?

(10) Summing up these points, what two forms do you
consider to be most similar?

(11) Draw transverse diagrams of a vertebrate, an
arthropod, and an earthworm, showing skeleton, body-
cavity, dorsal vessel, aorta, ventral vessel, heart, kidneys,
nervous system, appendages, etc., as far as you find them in
each.   Which two seem most alike?

(12) Can you better bring all three diagrams into harmony by turning any one wrong side up?   If so, what one
must be turned?

(13) Can you recall any such connection, in any vertebrate, between the dorsal and ventral blood-vessels, as you
find in the earthworm?   If so, where and what?

ANNELIDA (Segmented Worms).

The earthworm may be taken as a representative of this group, the members of which have a marked external segmentation of the body, but which lack jointed appendages. They also have a dorsal brain and a ventral ganglionic nerve-cord; the ganglia, like all other parts, being segmentally arranged. There are nephridia in each segment, while the circulatory system consists of a dorsal vessel in which the blood goes forward, a ventral vessel in which the flow is towards the tail, and segmentally arranged transverse vessels which connect the two. The annelids are divisible into several groups or orders, only two of which need be mentioned here.

Order I.—Chætopoda.

In these the body-cavity is well marked, as in the earthworm ; and each segment of the body bears setæ, which serve as locomotor organs. In some (Oligochætæ) the bristles are comparatively few, and they arise directly from the body-wall, while appendages of all kinds are lacking. A few of the Oligochætes live in the sea ; more occur in fresh water, but the great majority are terrestrial, and are familiarly known as " earthworms " or " angleworms," the latter name being given from their use in baiting fishhooks. The earthworms burrow in the soil, feeding upon decaying vegetable matter in the earth. They swallow earth and all, and come to the surface to deposit their well-known castings. In this way they work over the soil, and are of immense value to agriculture, as Darwin has shown in a most interesting volume on these lowly forms. Our earthworms are moderate in size, but in Africa, South America, and Australia giant earthworms, four to six feet in length and an inch in diameter, occur.

In other Chætopods (POLYCHÆTÆ) the bristles are numerous in each segment, and are borne on fleshy outgrowths (**parapodia**) from the sides of the body. The head bears fleshy feelers or tentacles, there are frequently horny jaws in the mouth, while eyes are commonly present —structures which are lacking in the Oligochætes. The Polychætes are largely marine, and occur in large numbers burrowing in the mud of the shores or sea-

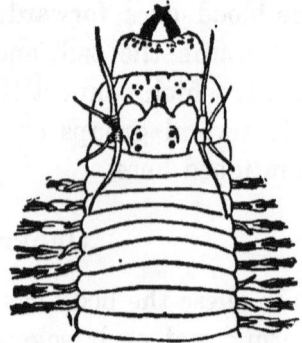

FIG. 112.—Anterior end of clam-worm (*Nereis*), showing jaws, eyes, tentacles, and bristle-bearing parapodia.

bottoms. Many of them are brightly colored, and marine worms are among the most beautiful objects in nature. They are largely carnivorous, and some of them are, to the associated life, terrible animals of prey.

ORDER II.—HIRUDINEI (Leeches).

The leeches have the body-segments ringed, so that one examining the outside would conclude that there were more segments than are really present. There are no bristles on the segments, but the hinder end always bears a sucking disc, while usually there is a second sucker around the mouth. The body-cavity is not distinct. There are two great groups of leeches—those with jaws around the mouth, and those which lack jaws.

The jawless leeches are aquatic, and occur in fresh water; more rarely in the sea. They live largely upon fishes, feeding upon the mucus covering the body. The jawed leeches have three jaws radiating from the mouth, and each jaw has its edge finely toothed. With these jaws they are able to cut the skin of vertebrates, upon the blood of which

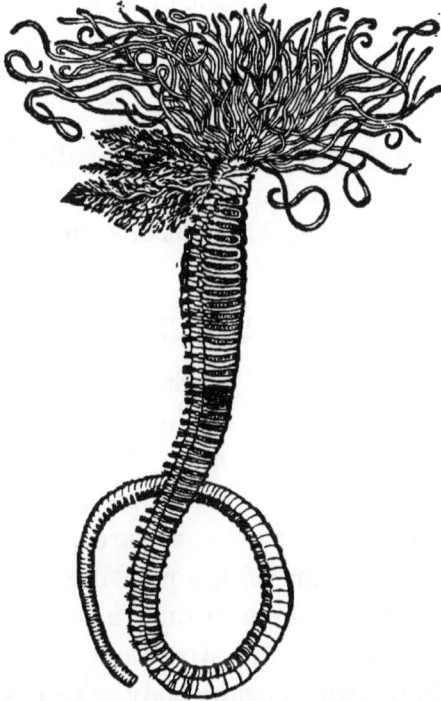

Fig. 113.—A tube-inhabiting Polychæte (*Amphitrite*). At the upper end are the tentacles, and just below to the left the gills.

they feed. This blood-sucking habit led to the use of leeches in medicine in those days when it was believed that if a man were sick his cure could be effected by still further weakening him. Most of the jawed leeches live in fresh water, but in the warmer parts of the Old World land leeches occur in the moist forests, and these form almost intolerable pests.

VERMES (Worms).

Under this heading are included a large number of forms commonly known as worms, but which are incapable of strict definition. In general it may be said that they have elongate bodies, without internal skeleton, without appendages, with a marked bilateral symmetry, and distinct dorsal and ventral surfaces. Further than this we can hardly go in a definition which will at once include all worms and at the same time not include other forms. Some of these worms are terrestrial, some aquatic, and some live as parasites on or in other animals. Omitting a number of microscopic forms and small groups, we may divide the Vermes into four classes: Plathelminthes or flat-worms, Nemathelminthes or round-worms, Annelids or segmented worms, and Molluscoidea.

Class I.—PLATHELMINTHES (Flat-worms).

In the flat-worms the body is flattened, is without appendages or skeleton; the mouth when present is on the ventral surface, and no vent occurs. There is no body-cavity aside from the digestive tract. Some are leaf-like, others are more elongate, and a very few are nearly cylindrical. The free-living and some of the parasites have an alimentary canal, but to this there is only a single opening, the mouth. Aside from the digestive cavity, the body is solid throughout, there being no such body-cavity as we have seen in all forms hitherto studied. The nervous system consists of a centre or "brain," always in the dorsal front portion of the body, from which nerve-cords run to various parts, there being usually two long cords which run backwards in a nearly parallel direction. Eyes may be present on the dorsal surface near the brain.

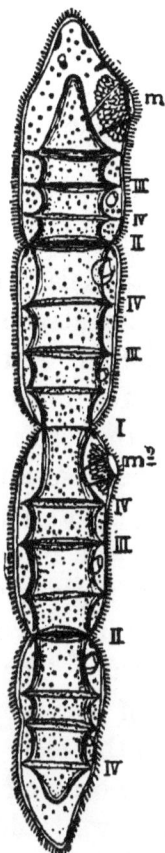

The capacity of reproduction by division is very well developed in these forms, especially in the non-parasitic groups. In these a second mouth will appear at about the middle of the body, then the body will constrict in front of the new mouth, and finally will divide into two worms. Not infrequently a new mouth will appear in each of the halves before the division is complete, so that we can have a chain of four or even eight animals connected together, and all the result of division of a single parent. Besides this reproduction by division, reproduction by means of eggs occurs. The Plathelminthes are divided into three orders—Turbellaria, Trematoda, and Cestoda.

ORDER I.—TURBELLARIA.

FIG. 114.—Process of division in *Microstomum.* After Graff. *m, m²*, mouths of successive generation; I-IV, successive planes of division.

These are small free-living forms which occur in fresh or salt water, and occasionally in moist earth. They are common in our ponds and streams, crawling over the bottoms or upon submerged sticks and stones. They have a mouth and digestive tract.

ORDER II.—TREMATODES.

Like the last, these have mouth and digestive tract, but they differ in being parasitic on or in other animals, and in having sucking discs (from one to many) developed upon the body. Some of them become serious pests. One form, the liver-fluke, produces the disease known as "liver-rot" in sheep. Other forms occur in man, especially in the tropics, being introduced in drinking-water. They cause serious sickness.

Order III.—Cestodes (Tapeworms).

The Cestodes are all parasitic in other animals. They differ from the Trematodes in the complete absence of mouth and digestive tract, since they absorb their nourishment through the skin. Usually they have ribbon-like bodies, and hence are commonly known as tapeworms. At

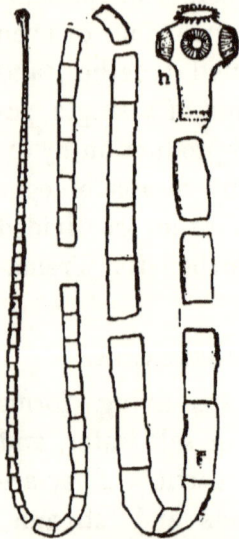

Fig. 115.—**Tapeworm** (*Tænia*) with proglottids from different regions of the body. *h*, head enlarged.

the anterior end are the means of attachment (hooks or suckers) by which the animal attaches itself to the lining of the intestine of its host, while usually the body becomes broken up into a series of joints or **proglottids.** There is continually a formation of new proglottids near the head, while the older proglottids, loaded with eggs, drop off and are carried out with the waste of the digestive tract. These tapeworms obtain entrance into the body in the food, man usually receiving his from raw or partially cooked beef or pork, and more rarely from fish. The proglottids and eggs, passing from the body, may fall where they may be eaten by cattle or swine. Inside their bodies, they undergo partial development in the muscles, and then when taken into the human body they complete their development.

Other vertebrates than man possess tapeworms. The cat gets hers from the mouse, the dog his from cattle and rabbits, the sharks from other fish, etc.

CLASS II.—NEMATHELMINTHES.

In these round-worms the body is long and cylindrical, and is covered with a firm cuticle. Usually both mouth and vent are present, but there is never any division of the body into segments. Some live freely in the water, some are parasitic in plants, and some infest animals. Among them are to be mentioned the vinegar and paste "eels," which are occasionally found in these substances. Here, too, belong the "horsehair-worms," which are frequently believed to be horsehairs converted into worms by soaking in water. These hairworms are at one period of their lives parasitic in insects, especially in grasshoppers. Some of the roundworms occur as parasites in man. The stomach-worms and pinworms of children belong to the round-worms, and these obtain entrance to the human system only as the exceedingly minute eggs are taken into the stomach by way of the mouth.

Worst of all the parasitic Nemathelminthes is the Trichina, which when adult is scarcely an eighth of an inch in length, and yet which not infrequently causes death. Man is usually infected with them by eating raw or partially cooked pork. In the pig they first appear in the alimentary canal, where the mothers bring forth myriads of living young. These young burrow

FIG. 116.—*Trichina*, encysted in human muscle. After Leuckart.

outwards into the muscles and there enclose themselves in a

capsule, where they remain indefinitely. If this infested flesh be eaten raw, the capsule is dissolved by the stomach, the young are soon born, and they in turn wander through the muscles, and, when numerous, this boring into the flesh causes severe sickness, and even death. The worst epidemic of this disease, known as trichinosis, on record occurred near Emmersleben, Saxony, in 1884. From one pig three hundred and sixty-four persons were infected, and of these fifty-seven died within a month. The moral which we have to learn from tapeworms and trichina is that our beef and pork should never be eaten raw, but should be cooked through.

Class III.—ANNELIDA (p. 239).

Class IV.—MOLLUSCOIDEA.

Under this heading are grouped a few forms, which in time past were considered as Molluscs (see p. 269), but which are now known to have only superficial resemblances to the clams, etc. There are two orders of these Molluscoids.

Order I.—Polyzoa (Moss Animals).

The Polyzoa are individually small, but by budding they form colonies of considerable size, the tentacles of the individuals giving the colony a mossy appearance. These tentacles surround the mouth in a more or less modified. circle, and by them the animals obtain their food. The body is sac-like, and the alimentary canal is bent upon itself so that the vent is near the mouth. Many of the colonies secrete an external skeleton, which may be horny or calcareous. Most of the Polyzoa are marine, but a few occur in fresh water.

ORDER II.—BRACHIOPODA (Lamp shells).

From the fact that the Brachiopoda possess a bivalve shell, these forms were formerly included among the molluscs near the clams. A little examination, however, shows that the resemblance between them is but slight. The two valves of the Brachiopod are unequal in size, and are dorsal and ventral, rather than right and left, as in the clams. Near the point where the two parts (valves) are

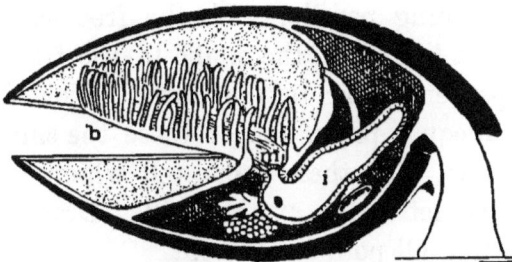

FIG. 117.—Diagram of a Brachiopod. *b*, tentacles around mouth, *m* ; *i*, intestine ; the shell black, the stalk to the right.

hinged together there is usually an opening * in the larger valve through which a fleshy peduncle or stalk projects, by means of which the animal is fastened to some support. Inside the valves, which can be closed by muscles, are the principal organs. Near the mouth are found a number of delicate tentacles (much like those of the Polyzoa), the disk which bears them being frequently rolled into a spiral. The alimentary canal is bent, but a vent is occasionally lacking.

The Brachiopods are all marine. There are few in existing seas; but they are among the oldest inhabitants, for the shells are found fossil in all rocks from the oldest down to the present time.

* In some the peduncle extends from between the valves instead of having a special opening.

THE CLAM: LABORATORY WORK.

For this purpose the student can use either the fresh-water clam or the long clam of the Northern sea-shore. For the study of the nervous system clams which have been a few days in alcohol are better than fresh specimens.

EXTERNAL.

Notice the shell; of how many parts or **valves** is it composed? Are the valves equal in size? They are joined by a **hinge**, dorsal in position, and each valve has a prominence (**umbo**) near the hinge. On each valve see the **lines of growth** running parallel with the free margin of the shell. Draw a line from the umbo to the free margin of the shell, perpendicular to the latter. This divides the valve into unequal parts, and of these the smaller is the anterior. Now with these facts tell which is the right and which the left valve of the shell. Draw one of the valves, inserting all points made out.

INTERNAL.

Remove the **left** valve from the clam by inserting a knife at either end close to the shell, and cutting the muscles which lie near the hinge-line. Then carefully remove the valve, seeing that all fleshy portions are left in the right valve. If properly done, this will leave the animal covered with a thin membrane, the **mantle**. Projecting through this, near the dorsal line, are the **adductor muscles**, which keep the shell closed, and which were cut in removing the valve. According to their position, these are known as the **anterior** and **posterior adductors**. Are the edges of the

248

mantle thickened ? Are the mantles of the right and left sides united anywhere along the free margin of the shell ?

Cut through the mantle near its ventral edge and fold back. Is it free back to the hinge line ? Cutting through the mantle opens the **mantle** or **branchial chamber.** In this several structures are to be noticed. Arising from the side of the body are plaited folds (how many ?), the **branchiæ** or **gills.** Are there branchiæ on the right side as well ? Extending downward between the gills is the soft **abdomen,** terminated at the anterior ventral angle by a more solid **foot.** In front, just ventral to the anterior adductor, are two pairs of fleshy flaps, the **labial palpi,** and where they meet at their junction with the body is the **mouth.** At the posterior end of the animal look for two fleshy tubes **(siphons)** formed by the edge of the mantle.* Run a wire in each from the outer end, and see where it appears inside the shell. The ventral siphon is the incurrent or branchial siphon; the dorsal is the excurrent or cloacal siphon. Draw the parts so far made out.

Just beneath and behind the hinge is the heart, its position in the living animal being readily seen by its pulsations. Carefully cut into the chamber in which it is situated and make out a central **ventricle,** rather dense in texture, and leading to it on either side a delicate tubular **auricle** which brings the blood from the gills to the ventricle. Notice the intestine passing through the ventricle. Just in front of the posterior adductor is the dark **organ of Bojanus** or kidney. Draw the parts made out.

The alimentary canal and the nervous system are best followed in specimens which have been in alcohol a few days. In such a specimen insert a probe into the excur-

---

* These are small in the fresh-water clams, but are greatly developed, and form the part commonly but erroneously called the " head," in the long clam.

rent siphon.   Notice that it does not enter the branchial chamber.   Cut through the thin membrane between the gills of the right and left sides, posterior to the abdomen. This lays open the **cloacal chamber** into which the probe extends.   In the dorsal wall of this chamber, just below the posterior adductor, see a pinkish or orange body, the **parieto-splanchnic ganglia.**   From this trace backward nerves which curve forwards along the base of the gills. Also trace two nerves forward, one on either side of the body, until they meet in a pair of **cerebral ganglia** just above the mouth.   Are the two cerebral ganglia connected directly with each other?   From the cerebral ganglia trace a pair of nerves downward to the **pedal ganglia** lying between the abdomen and the foot.   Sketch the nervous system.

Beginning with the intestine where it leaves the heart, trace it posteriorly.   On which side of the posterior adductor does it pass?   Where does it empty?   Trace it forward from the heart, carefully picking away the surrounding tissue with the needles, into and through the abdominal mass, and plot the coils which it makes.   It will be found to pass into a rather large saccular **stomach,** on either side of which is the dark-green **liver.***   Trace the œsophagus from the stomach to the mouth.

Take a clam which has been hardened for a couple of weeks in strong alcohol or formol.   Cut it transversely in slices a quarter of an inch thick, using a sharp scalpel for the purpose.   Draw the sections and name all the parts found.   This can be done easily if the previous dissection has been intelligently done.

* In a pocket of the stomach in the long clam will be found a structure of unknown function, the **crystalline style,** transparent, an inch or more in length.

THE OYSTER.

Oysters in the shell should be used. Find the hinge as in the clam. Do you find lines of growth? In the same way as in the clam distinguish anterior and posterior, right and left valves. Is the right or the left valve convex?

Break the shell at the hinder end and, inserting a knife, cut the adductor muscle so as to remove the left valve.* How many adductors do you find? Is the mantle edge thickened and united as in the clam? Do you find any siphons? What other peculiarities do you find in the edge of the mantle?

Remove the mantle from the left side and trace the parts. How does the foot compare with that of the clam? How do the palpi differ? How many gills? Which adductor— anterior or posterior—is absent? Find the heart, just in front of the adductor. Lay open the pericardium. How many auricles and how many ventricles are present? Trace the alimentary canal through the body from the mouth to the vent. How is it related to the heart?

* If you do not know where the adductor is, study a shell already removed and find the scar made by it.

SQUID: LABORATORY WORK.

EXTERNAL FORM.

The head, separated from the body by a "neck," bears at its anterior end a circle of **tentacles**; how many? Are all of these of equal length? If not, which pair is the longer? On the side of the head are the eyes; behind the eye is a fold of the skin, the **olfactory organ**. The body is surrounded with a **mantle**, bearing at the posterior end a pair of large **fins**. Is the mantle joined to the body all around? If not, where is it attached? Projecting from the mantle opening is the end of a fleshy tube, the **siphon**. The side of the body on which the siphon occurs is usually called the **ventral** side.

Sketch the squid from the side, showing these points, not omitting the color spots (**chromatophores**).

Examine the tentacles more carefully. On their inner surfaces see the stalked **suckers**. Are they similarly arranged on all the arms? Examine a sucker with the hand-lens, making out the fleshy lip, the horny hooks, and a fleshy bottom (**piston**) in the central cavity. Sketch a sucker, considerably enlarged.

INTERNAL STRUCTURE.

Place the squid in the dissecting-pan, siphon uppermost. Cut the mantle longitudinally a little to one side of the middle, beginning at the free edge and carrying the incision to the end of the body. This lays open the **mantle chamber**.

252

Lift the cut edges carefully, looking for the **median mantle artery** running from the body to the mantle. Pin out the mantle and make out the following points:

The siphon; notice its inner end; just behind it is the end of the intestine. On either side of the siphon are the **siphonal cartilages**, grooved on the surface. Look on the edge of the mantle and find a ridge. Close up the mantle and see how the parts interlock.

Behind the siphon, at either side of the body, are the **gills**. What structure have they? Can you see any vessels connected with them? Follow the intestine back from the vent. Is it free, or is it tied down to the underlying structures? Notice that it passes across a dark-colored sac—the **ink-sac**. Some distance behind the gills see a vessel, the **post-cava**, coming from the side of the mantle forward to the body.

The other features vary considerably accordingly as the specimen is male or female. In the female the hinder part of the body is occupied with eggs, while upon that part between the gills is the large transversely striated **nidamental gland.*** When these are carefully removed the structures are much as in the male.

On either side of the intestine, a little behind the ink-sac, is the small opening of the kidney; the **kidneys** themselves stretch back behind the base of the gills. They are irregular in shape. When they are removed† there will be seen in the median line the **systemic heart**. Behind, it gives off an arterial trunk, which soon divides to form the median mantle artery already noticed, and the **lateral mantle arteries** which follow the postcavæ. On either side it

---

* These secrete the capsules in which the masses of eggs are laid.

† Cut through the thin wall of the kidney just behind the gill, pull off the thin skin, and wash away the granular contents.

receives a **branchial vein,** coming from the gill; while in front it gives off an anterior aorta which runs forward.

Look on the side of the gill nearest the mantle and see the **branchial artery.** Trace it towards the middle line and find the **branchial heart,** just behind the branchial vein. This receives the blood from the postcavæ already noticed, and also from a precava which comes from in front through the kidney, but is not so easily traced. The course of the circulation may be briefly described as follows: The blood is forced to all parts of the body by the systemic heart. After supplying these regions it collects in the pre- and postcavas and is brought to the branchial hearts, which pump it through the branchial arteries to the gills. From the gills it returns to the systemic heart by way of the branchial vein to repeat its circuit. Sketch all parts made out.

Carefully trace the intestine backwards from the vent, removing the systemic heart and the remains of the kidneys. Just behind the level of the systemic heart it will be found to enter the thick-walled, muscular **stomach.** This stomach gives off, behind, a large, thin-walled blind sac, which extends far back into the body mass. Close to where the intestine leaves the stomach the **œsophagus** enters it. Trace the œsophagus forward to the region of the neck, *but not farther at present.* In its course it can be followed through the **liver.** Sketch the alimentary tract as if viewed from the side, inserting intestine, ink-sac, stomach, blind sac, liver, and œsophagus, leaving room for the anterior end of the latter to be inserted later.

With a single stroke of a sharp scalpel split the head longitudinally, making the cut as nearly as possible in the median plane. In the section thus made the anterior end of the alimentary tract and the central part of the nervous system can be easily studied.

Just inside the mouth, which is placed in the centre of the circle of arms, is the oval **buccal mass**, which is only slightly connected with the rest of the head. In this find the two horny jaws, black at the tips, and shaped something like the beak of a parrot. Do these jaws work in a vertical or in a horizontal plane? The cavity of the mouth lies inside these jaws and passes nearer to the dorsal jaw. Just inside the mouth-cavity is a pocket given off on the ventral side, in which will be found a horny **lingual ribbon**, covered with minute horny teeth. Could this ribbon be used in rasping the food after it had passed the jaws? Notice that the bulk of the buccal mass is made up of muscles arranged to move jaws and lingual ribbon.

From the buccal mass trace the œsophagus backward to the point where it was left in the previous dissection. Do not cut at first in tracing it, as you would be apt to injure other portions. If the section of the head be in the median plane, the course of the œsophagus will be easily followed without dissection. If not, it can be traced later after the nervous structures have been studied.

A little back of the buccal mass some harder, cartilage-like structures will be seen in the cut surface of the head. These form a **brain capsule**, resembling in some respects the vertebrate skull. In the dorsal side of this will be found a large centre, the **cerebral ganglion**, while on the ventral side two somewhat smaller ganglia occur. The anterior of these is the **pedal ganglion**, and from it nerves can be traced running into the arms. The posterior is the **visceral ganglion**. Between the cerebral on the one hand and the pedal and visceral ganglia on the other passes the œsophagus. In one half of the head demonstrate by dissection that these ganglia are connected. Except that the ganglia are much closer together and the connections correspondingly shortened, are the relations the same as in the clam?

Just ventral to the visceral ganglion is an enlargement of the cerebral capsule; this is the ear. Cut into this and notice that it has an irregular cavity. Is there a similar structure on the other side of the head? Sketch the section of the head, showing the ganglia, jaws, lingual ribbon, œsophagus, and ear, in the drawing already made of the alimentary tract.

Split one half of the head in a horizontal plane, having the section pass through the middle of the eye. In the section thus made study first the eye itself. This is covered externally with a transparent **cornea,** and inside contains two chambers, separated from each other by the solid **lens.** The outer chamber in turn is partially divided by a circular fold, the **iris.** The inner chamber is bounded internally by the **retina,** the outer surface of which is marked by a thin layer of black pigment. Behind and dorsal to the eye is the **optic ganglion,** bounded posteriorly by a cartilage wall. Trace the connections of the optic and cerebral ganglia.

Cut into the dorsal region of the mantle from the outside and find the horny **pen.** Continue the cutting so that it may be taken out. Sketch it.

COMPARISONS.

With two columns, one for oyster and clam and one for squid, answer the following questions:

(1) Is there a distinct head?

(2) Are there cephalic tentacles?

(3) Is there a bivalve shell?

(4) Is the siphon, if present, a part of the mantle ?

(5) Did you find any eyes ?

(6) Are adductor muscles present ?

(7) Is there a bivalve shell ?

(8) Are the gills leaf-like or plume-like ?

(9) Are there jaws ?

(10) Is there a lingual ribbon ?

(11) Are there branchial and systemic hearts ?

(12) Is there an ink-sac ?

ACEPHALA.

In the Acephala, as the name implies, there is no distinct head. The body is flattened from side to side, and the two sides are almost exact repetitions of each other. On either side of the body there is a strong outgrowth of the body wall, the **mantle**, which secretes on its outer surface the **shell**, which is divided in the median line so that two halves or valves result. Between the mantle folds and the body is the mantle-chamber, and into this on either side there usually hangs down a pair of leaf-like gills.* From the lower surface of the body projects a muscular **foot**. With these features the animal presents a marked resemblance to a book in which the valves represent the covers; the mantle, gills, body, and foot, seven leaves.

Where the two valves are hinged together there is an elastic ligament which tends constantly to open the valves, which are closed by means of adductor muscles extending from one valve to the other. Usually there are two of these muscles—anterior and posterior, but the anterior of these may disappear.

In some, as in the oyster, the mantle edges are free from each other throughout their extent; but not infrequently they become fused in places, leaving openings between. At the posterior end this fusion frequently results in the formation of two tubes or **siphons** connecting the outer

* It is not necessary here to include the gill features of *Cuspidaria, Silenia*, etc.

world with the mantle-chamber. When these siphons
become greatly developed there are connected with them
strong retractor muscles, to
draw them back at times of
danger, etc. All of these
muscles—adductors, retractors, etc.—leave their impress
on the shell, so that the student, with the shell alone,
knows of some of the structures of the soft parts.

Water is drawn into the
mantle-cavity by means of

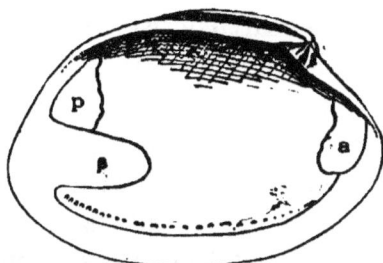

Fig. 118 —Inside of bivalve shell
showing muscular impressions. *a,*
anterior adductor; *p,* posterior adductor; *s,* siphonal muscle.

very minute hair-like structures (cilia) which cover the
gills and other parts. These cilia are in constant motion,*
and thus currents of water are produced, flowing always in
one direction. This water brings oxygen to the gills and,
through them, to the blood. It also brings minute animals
and plants. These are passed on to the labial palpi, which
are similarly covered with cilia, and from these organs the
cilia force the food into the mouth.

In the nervous system we always find cerebral, pedal, and
visceral ganglia, the first being above, the others beside
or below, the alimentary canal. Ears are present, connected
with the pedal ganglia; and eyes may be present, either
upon the edges of the mantle or at the tips of the siphons.

The alimentary canal is always provided with stomach
and liver. Connected with the stomach a blind sac frequently occurs, and in this there may be a peculiar transparent rod, the **crystalline** style, of uncertain use. The

* The teacher should demonstrate this ciliary action under the
compound microscope.

intestine goes from the stomach first towards the foot, then mounts towards the hinge-line, and frequently passes through the ventricle of the heart.

The heart consists of a single ventricle and usually two auricles, but sometimes there is but one of the latter. The heart is situated in a chamber (**pericardium**), which is connected by means of a pair of convoluted kidney tubules (**organ of Bojanus**) with the exterior.

A thoroughly satisfactory classification of the Acephala has not yet been worked out. Possibly the best is that based upon the structure of the gills, but a more convenient one for our purposes is based upon the presence or absence of a siphon.

Order I.—Asiphonida.

The edges of the mantle free; no siphon present. Most prominent of this order are the oysters. These are all marine, species being found in all but the colder seas. In these forms the animal lies upon one side, and there results an inequality of the valves. On our east coasts oysters extend from the Gulf of Mexico to Cape Cod. Further north (except in the Bay of Chaleur) they are not found native, but are "planted." The centre of the oyster industry is Baltimore. In 1894 the oyster-fishery of the United States amounted to over $16,000,000.

In the scallops the shell is fluted, and the valves may be unequal or similar in shape. These molluscs can swim freely by rapidly opening and closing the valves of the shell; and they are further noticeable from the fact that around the edge of the mantle are a series of rather complicated eyes. The "scallops" of the markets are the adductor muscles of these molluscs. In the pearl-oysters the inner

layer of the shell has a pearly appearance, and these forms
also produce, like some other molluscs, the precious pearls.
These pearls are really the shell-forming secretions of the

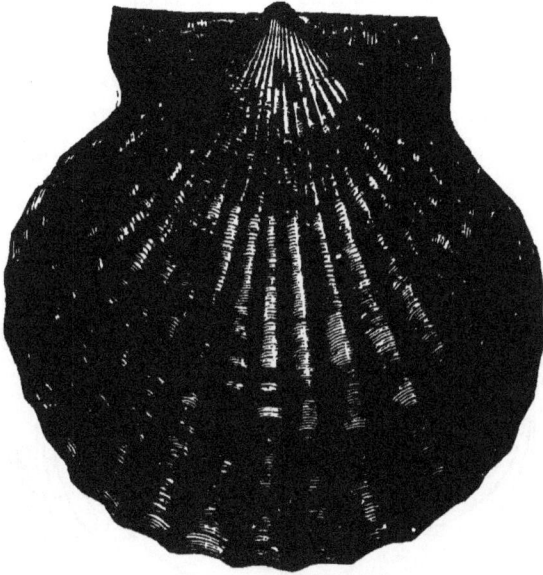

FIG. 119.—Scallop (*Pecten irradians*). From Binney's Gould.

mollusc around some foreign body, and they receive their
beauty from the way in which the shell is deposited around

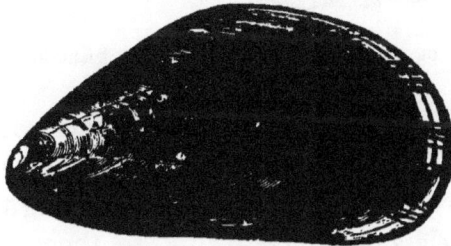

FIG. 120.—Salt-water mussel (*Mytilus edulis*).

the centre. Fresh-water mussels, to be referred to a few
lines below, also form pearls of value. The shell of the
pearl-oyster also has its value, for it furnishes the mother-

of-pearl used for knife-handles, for inlaying, etc. The pearl-oysters occur in the Indian Ocean, and also in the Bay of Panama.

The salt-water mussels, so abundant on the mud flats all along Northern shores have a peculiar gland in the foot which secretes strong silky threads (**byssus**) by which these animals anchor themselves. The common species, which occurs both in Europe and New England, is called the edible

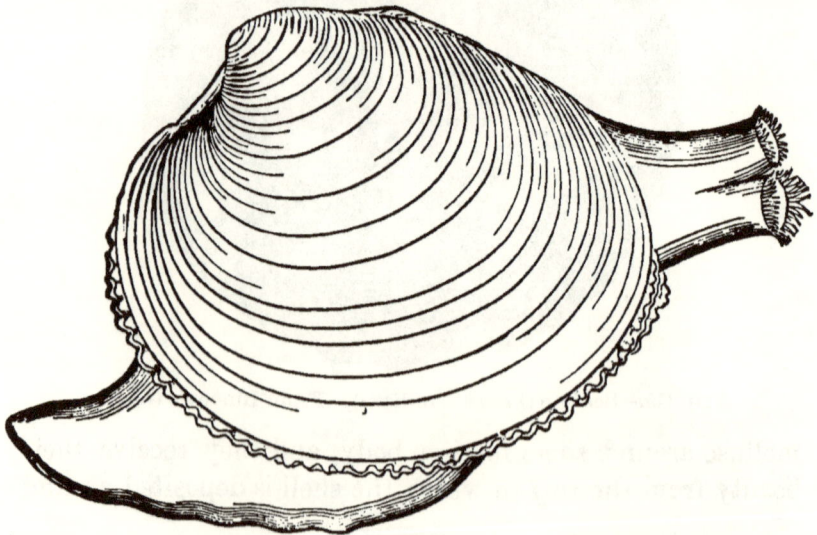

FIG. 121.—Quahog (*Venus mercenaria*), with foot and siphons extended.

mussel; but not infrequently severe sickness follows its use as food. The fresh-water mussels are especially abundant in America, the Mississippi basin being their centre. They are useless as food, owing to their strong taste. There are possibly a hundred species of these forms in America; over six hundred so-called species have been described. In their siphonal structure they form a transition to the next group.

ORDER II.—SIPHONATA.

In these the margins of the mantle have grown together posteriorly into a double tube or siphon, and accordingly as this siphon is developed the animal can burrow below the surface and still obtain its necessary supplies of water and food; for these tubes can reach the surface, and through them there is a continual flow of water—inward through the ventral, outwards through the dorsal, passage.

The great majority of bivalve molluscs belong here, but there are comparatively few of general interest. The largest of all clams, the giant clam of the East Indies, with shells sometimes weighing over 300 pounds, belongs here, as do the quahog and the long clam, which are used as food. One of these forms, the Teredo or ship-worm, is a serious pest, as it bores in wood, destroying the piles of wharves, the bottoms of boats, etc. Their burrows run to long distances, but all their food and water must be drawn in through the siphons. One great inundation in Holland at the beginning of the last century was directly due to the borings of these forms.

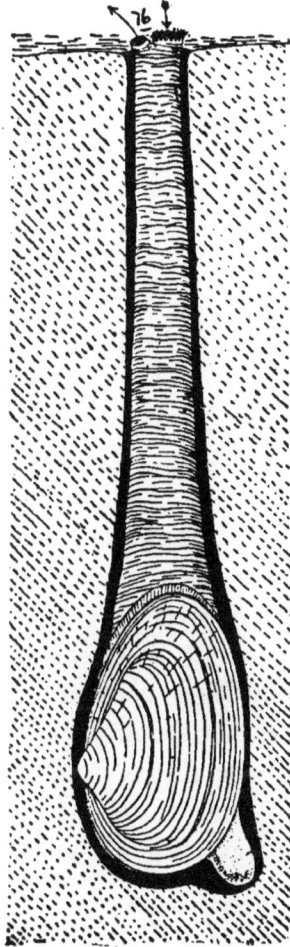

FIG. 122.—Long clam (*Mya arenaria*) buried in the mud. The arrows show the currents in the siphons.

CEPHALOPODA (SQUID AND CUTTLEFISH).

The Cephalopods derive their name from the fact that the circle of tentacles or arms around the mouth (i.e., on the head) was compared to the foot of other molluscs. Later investigations show that these tentacles represent but a part of the foot, the siphon also belonging to the same category. These same arms, which are either eight or ten in number, bear sucking organs by means of which these animals hold fast their prey. In only the pearly nautilus are the arms lacking, and here they are replaced by about a hundred smaller organs.

The head, which is separated from the body by a distinct neck, bears a pair of eyes—simple in the nautilus, but almost as complex as those of man in the other forms. In these more highly developed eyes there is retina, lens, iris, cornea, and cavities resembling those occupied by the aqueous and vitreous humors. Yet the resemblances are superficial; the structures are in reality totally different.

The mantle is connected with the body in the region of the so-called back. Below, it encloses a good-sized mantle-cavity, open in front. It is very muscular, and the opening about the neck can be closed at will, so that the only connection between the mantle-chamber and the outside world is through the tube of the siphon. If one of these animals fill its mantle with water, close the neck opening, and then force out the water by contracting the mantle, the water will stream from the siphon in a strong jet, which by its reaction forces the animal in the other direction. This apparatus forms with many, and especially with the squid, the chief organ of locomotion, and in these the tip of the siphon can be bent in any direction, so that the animal may go forwards, backwards, etc., according as it wishes.

In the mantle-cavity are one or two (Nautilus) pairs of feather-like gills, and into the same chamber empty the ducts of the kidneys and reproductive organs, as well as the intestine, and the ink-sac connected with it. This last organ secretes a dark-colored fluid, which when discharged into the water makes a cloud, and thus the animal is enabled to escape unseen. From this ink the pigment sepia and some kinds of India-ink are manufactured.

Imbedded in the skin of the mantle are pigment spots or **chromatophores**, which are interesting from the fact that they can be enlarged or contracted by the nervous system. When enlarged they nearly touch each other, and thus give the body their general hue (red). When contracted they appear as minute black points, while the general body color (translucent white) then prevails. As a result we have in these animals a power of color-change far more striking than that of the chamæleons.

Most living Cephalopods have no external shell. Inside of the back, however, is a shell—the pen—which may be either feather-shaped and horny, or broader, thicker, and calcareous. In this last condition it furnishes the "cuttlebone" so often given to cage-birds. The paper nautilus has a shell which is formed only by the female; it is secreted, not by the mantle, but by one pair of the arms, and this shell is really a protection for the eggs. In the pearly nautilus, on the other hand, there is a true shell, which is coiled in a spiral and is divided by partitions into a series of chambers, only the outer one being occupied by the animal. Similar chambered shells are very abundant among fossils.

The mouth is armed with a pair of horny jaws shaped much like those of a parrot. These are very efficient in biting food; but any morsels taken into the mouth are subjected to further subdivision by means of the lingual ribbon, which is, as its name implies, a ribbon-like membrane, bearing on its surface numbers of minute teeth, which rasp the food into fine shreds.

The heart is situated in a pericardium and is systemic; that is, it pumps the blood returning from the gills to the various parts of the body. A peculiarity of the circulatory system is that in all, except the pearly nautilus, the vessel carrying blood to the gills develops a special pumping organ, the branchial heart.

The various ganglia of the nervous system are (except the stellate ganglia) placed close together in the head, and from this centre nerves radiate to all parts of the body, those going to the tentacles being connected with each other by a circular cord.

The Cephalopods are all marine. They are carnivorous, feeding upon fishes, etc., which they capture with their arms and hold fast by their numerous suckers. The larger forms might be no mean antagonist for man; but the monster described by Victor Hugo is without counterpart in nature. The Cephalopods are divided into two orders, according to the number of gills.

ORDER I.– TETRABRANCHIATA.

In the Tetrabranchs there are two pairs of gills (*i.e.*, four in all); the head bears numerous short tentacles without suckers, and the body is enclosed in a chambered shell.

The pearly nautilus is the only living representative of this group. It occurs in the East Indian seas, and, while the shells are very common, the animal is very rare in museums. In geological times allied forms were very abundant, and are known as Ammonites (with tightly coiled shells), and Orthoceratites (with straight shells), etc.

ORDER II.—DIBRANCHIATA.

These have two gills (one pair), and long, sucker-bearing arms. An ink-sac is always present. The order is subdivided into the OCTOPODA, in which there are eight arms,

FIG. 123.—*Octopus bairdii.* From Verrill. One arm on the right side is modified for purposes of reproduction.

and the DECAPODA, in which the number is increased to ten by the addition of a pair of longer arms. In the Octopoda there is no internal shell, and the body is saccular. Here belong the octopus, poulpes, etc., as well as the paper nautilus, which does not sail with its shell as a boat, and its broadened arms erect to catch the wind, as it is often said to do. The Decapoda include the squid, the sepia, and other forms. The smaller squid are abundant, and are caught in large numbers for bait in fishing for cod. Near Newfoundland, and in other parts of the world, giant squid are occasionally found, the largest one known having a body length of twenty feet. The length of the arms was not mentioned in the account.

COMPARISONS.

With two columns, as before, for clam, oyster, and squid, answer the following questions :

(1) Is the body bilaterally symmetrical ?

(2) Is there a mantle ?

(3) Are gills present ?

(4) Is there a foot ?

(5) Do you find cerebral, pedal, and visceral ganglia ?

(6) Does the alimentary canal pass through the nervous system ?

MOLLUSCA.

Oysters, Clams, Snails, and Cuttlefish may be taken as examples of the ten thousand different species which are known as Molluscs. The name comes from the Latin *mollis*, soft, and alludes to the fact that, aside from the shell, the body has no conspicuous hard parts. This, however, is a point of no real importance in classifying animals.

Molluscs vary greatly in appearance; but if we carefully compare the points which all possess in common, we can construct an ideal mollusc, from which any form may be derived by additions here and modifications there. Such a typical mollusc is described below.

The body is saccular, and bilaterally symmetrical. There

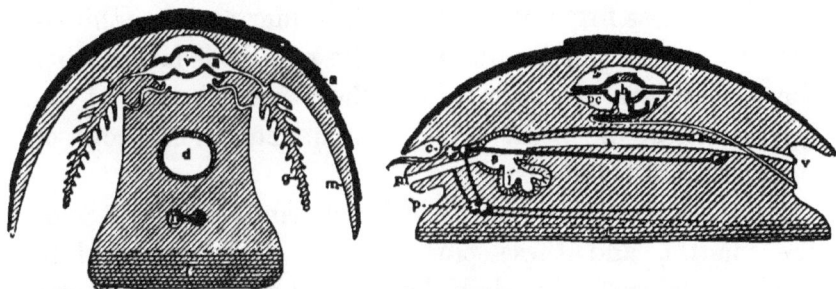

Fig. 124.—Transverse and longitudinal sections of a schematic Mollusc. *a*, auricle; *c*, cerebral ganglion; *d*, digestive tract; *f*, foot; *g*, gill; *h*, heart; *i*, intestine; *l*, liver; *m*, mouth; *n*, nervous system; *p*, pedal ganglia; *pc*, pericardium; *s*, stomach; *v*, vent.

is, above, a conical **visceral mass**; below, a muscular **foot**; while from either side a fold of the body-wall extends outwards and downwards as a **mantle.** Between the mantle

269

and the body and foot is a **mantle chamber,** or, since it frequently contains the gills (branchiæ), it is frequently called the **branchial chamber.**

The outer surface of the mantle and the dorsal part of the body frequently have the power of secreting a **shell** composed, chiefly, of carbonate of lime. This shell in some forms becomes split along the median line, so that two halves or **valves** result. In most other forms the shell becomes coiled into a spiral, and when this occurs the primitive symmetry becomes lost in part.

Shells increase in size during the life of the animal. The mantle is continually laying down new layers of shell inside of those first formed, hence the older parts are thicker than the newer portions. Then the mantle is larger when the new layers are secreted, so these project beyond the layers outside of them. As a consequence there occur on the outside **lines of growth.**

In many species there are colored bands or spots upon the mantle, and these parts secrete carbonate of lime similarly colored, the result being that the shell is correspondingly striped or spotted. Again, in some, the edge of the mantle is produced into finger-like lobes, etc., and these produce spines and the like upon the shell.

Shells are frequently spoken of as the houses or homes in which the animals live. As will be seen from the above, the shells are as much a part of the animal as is the carapax of a lobster or the wings of a butterfly. The oyster or snail can never leave its shell.

In most molluscs folds of the skin extend from the body-wall into the mantle-chamber. These are the **branchiæ** or **gills.** Inside of them are blood-vessels, and through their thin walls the blood is brought into close connection with the oxygen dissolved in the water, just as is the case in the gills of a fish. In the common terrestrial molluscs gills are absent, but the inside of the mantle-chamber is lined with a fine network of blood-vessels, so that the whole organ resembles somewhat a **lung**, and has received that name.

In the flow of the blood there is a great difference between the mollusc and the fish. In the mollusc the blood returns at once from the gill to the heart, and is then forced by this organ to all parts of the body. The heart is situated in a chamber or **pericardium**, and consists of one or two (right and left) **auricles** which receive the blood, and a **ventricle** which pumps it to the body. In the squid accessory or **branchial** hearts are added. These are placed at the bases of the gills and force the blood through these organs, from which they return to the other or **systemic heart**, to go to all parts of the body.

In all molluscs except the Acephala the region of the mouth is provided with a **lingual ribbon**. This is a band of horny material, bearing on its free surface rows of hard and sharp teeth, so that the whole resembles a flexible file. It is supported in such a way that it may be moved back and forth, thus rasping the food. In some Gasteropods it can even be used in boring holes in the shells of other molluscs. This lingual ribbon is constantly growing at its deeper end, so that the loss by wear in front is continually made good. The teeth on the ribbon vary in number and shape in different species. In some there are but three in a transverse row, while in others there may be over one hundred.

In the ideal mollusc the alimentary canal goes straight through the body from mouth to vent. In nature it usually has some convolutions, increasing the amount of digestive surface. In the Cephalopods and in most Gasteropods it becomes bent on itself, so that the vent is far in front, either upon the right side or even in the median line.   In the Gasteropods, when it is median, it is close to and dorsal to the mouth.   In the Cephalopods it is ventral.

The nervous system consists of at least three pairs of ganglia and the cords or commissures connecting them, as well as the nerves going to the various parts.   These ganglia are the **cerebral,** above the mouth; the **pedal,** primarily in the foot; and the **visceral,** farther back in the body.   Both pedal and visceral ganglia are below the intestine; the pedal supplying the foot, the visceral the body and the mantle.   To these three pairs others are frequently added.   Sometimes the ganglia are widely separated, when the commissures are correspondingly lengthened; or they may be brought close together, with shortened connecting cords.

Some molluscs lack organs of special sense; others have eyes and ears.   The ears are little sacs, usually near the pedal ganglion, but the eyes may have various positions. They may be on the sides of the head (squid), or on the sides or tips of tentacles arising from the head (snails), or scattered over the back (some slugs and chitons), or on the edges of the mantle (scallops), or on the end of the siphon (some clams).   In some they are merely spots which have the power to distinguish between light and darkness, and from these all degrees of development may be found to the extreme in the squid, where these organs are scarcely inferior to those of vertebrates in structure.

For kidneys the molluscs have one or two organs consisting of convoluted tubes opening at their inner end into the pericardium and communicating with the exterior at the other.

In some the sexes are separate; in others, like our land

snails, they are united in the same individual. All molluscs, with very few exceptions, lay eggs, from which the next generation is produced.

Molluscs are divided in different ways by different authorities. For our purposes we may recognize four divisions or classes: Placophora, Gasteropoda, Acephala, and Cephalopoda.

Class I.—PLACOPHORA.

Here belong a few forms known as Chitons. They are separated from all other molluscs by many points of internal structure, while externally they may always be recognized by having a dorsal shell composed of eight transverse plates, which overlap from in front backwards, like the shingles on a roof. All are marine.

Class II.—GASTEROPODA.

The Gasteropods receive their name from the fact that the foot usually forms a large sole or creeping disc extending along the ventral side of the body. There is a distinct head, which usually bears sensory tentacles, and the eyes are commonly placed at the bases or on the tips of one pair of these structures. In some cases, as in most land-snails, these tentacles can be pulled back into the body.

In the majority of forms gills are developed in the mantle-chamber. In a few there is a pair of these organs, but in many one gill disappears, while in other species both true gills entirely disappear, and are either replaced by secondary gills developed on the back or in other regions; or the mantle-chamber may be richly lined with blood-vessels and thus be converted into an organ (**lung**) for breathing air. This is the case in all of our common land-snails.

In all Gasteropods a shell is present in the young, but in many it is lost before the animal becomes adult. It is never a bivalve structure, but is either plate-like or is coiled in a spiral. In some the spiral is flat, in others it may be elongate, and the turns may be either to the right or to the left, right-handed shells being in the great majority. In a large number of Gasteropods a shell-like structure (**operculum**) is developed on the dorsal surface of the hinder part of the foot, and when the animal withdraws itself into the shell this operculum closes the opening after all the soft parts are inside.

Some of the peculiarities of the nervous system form the basis of the subdivision of the Gasteropods. In one group (Euthyneura) the ganglia and the cords connecting them are much as in the clam. In the other (Streptoneura) the cords leading back from the brain become crossed so that the nerve which starts from the right side goes to a ganglion on the left, and *vice versa*.

In all Gasteropods a lingual ribbon (p. 270) is present, and this works against a plate or " jaw " on the upper side of the mouth. The alimentary canal is rarely straight. Usually there are convolutions, and the whole is so bent upon itself that the vent is carried far forward, and may be placed upon the neck just above the mouth. Sometimes it, or the liver connected with it, become greatly branched.

Subclass I.—Streptoneura.

In these the nervous system is twisted; there is but a single pair of tentacles upon the head; and the gills are placed in front of the heart, a condition which leads many naturalists to call the group " Prosobranchs."

ORDER I.—DIOTOCARDIA.

In these forms the body retains its bilateral symmetry to a considerable degree, and externally may appear perfectly symmetrical. The name implies the existence of two auricles to the heart. In the limpets the shell is a flattened cone; in the abalones it is somewhat ear-shaped and very weakly spiral, but in the top shells it is strongly spiral. The abalones alone have any economic value. Their shells, remarkable for having a series of holes in them, are composed of a greenish mother-of-pearl, which is extensively used in inlaid work.

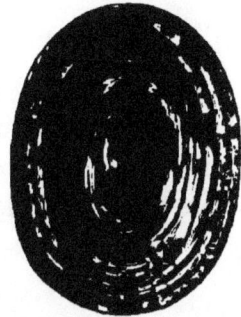

FIG. 125.—Limpet (*Acmœa testudinalis*). From Binney's Gould.

ORDER II.—MONOTOCARDIA.

Here belong the great majority of marine snails, all of which agree in having but a single gill and a single auricle to the heart. Few of them have any economic interest aside from those which feed upon oysters and other valuable shellfish. These injurious forms—commonly known as "drills"—are able to bore holes through the shells of oysters, etc., by means of their lingual ribbons. Many, however, are great favorites with collectors, among them the strombs, cones, cowries, and olives. Some of the cones are noticeable from the fact that they have a poison-gland connected with the lingual ribbon. Some species formerly grouped as a

FIG. 126. — Stromb (*Strombus pugilis*). After Woodward.

distinct order of Heteropoda are especially modified for a life on the high seas.

Subclass II.—Euthyneura.

In the Euthyneura the nervous system is without a twist, and the head almost always bears two pairs of tentacles.

Order I.—Opisthobranchia.

These forms are all marine, and have but two divisions to the heart—an auricle and a ventricle, the latter being in front of the former.   Some are provided with a spiral shell, while others—called Nudibranchs or naked molluscs—are

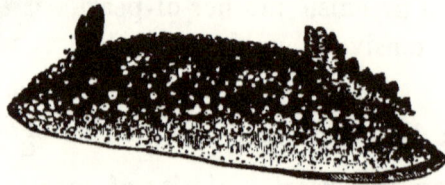

Fig. 127.—Naked mollusc (*Doris*), showing the gills, above to the right.

without such protection.   In the nudibranchs there are commonly developed gills upon the dorsal surface, and in the living condition these forms are, from their bright colors among the most attractive of molluscs.   Here, too, are forms (Pteropods) especially developed for a life on the surface of the ocean, the foot being modified into a pair of wing-like structures.

Order II.—Pulmonata.

The great majority of the land and fresh-water snails and slugs belong here.   In them gills have disappeared, and the mantle-cavity has been modified into an organ (lung) for breathing air, the opening to which is to be seen on the right side of the body. Over six thousand species belong here, some (snails) having a well-developed spiral shell, while the slugs are apparently shell-less; but in these slugs one can frequently find a rudimentary shell imbedded in the mantle.

CLASS III.—SCAPHOPODA (Tooth-shells).

In these the mantle edges are fused below, forming a tube, and as a result there is formed a tubular shell, open at both ends, in shape something like the tusk of an elephant. The foot is large, and adapted for digging in the sand, in which these animals live. There is no distinct head, but the mouth is provided with a lingual ribbon. In the anterior part of the mantle-cavity are a pair of bunches of long threads of unknown function; possibly they are sensory, possibly respiratory, in nature. All of the tooth-shells are marine.

CLASS IV.—CEPHALOPODA (see p. 264).

CLASS V.—ACEPHALA (see p. 257).

STARFISH: LABORATORY WORK.

EXTERNAL.

The body is shaped like a five-rayed star; in it distinguish the central **disc** and the arms or **rays**. In the centre of the disc find the **mouth**. The side on which it occurs is called the **oral surface**. Running along the oral surface of each arm are the fleshy **tube-feet or ambulacra**, and the regions of the oral surface in which they occur are known as the **ambulacral areas**. Sketch this surface in outline, showing the parts.

The surface opposite the mouth is the **aboral surface**. Does it have ambulacra? By feeling and bending see that this surface is composed of numerous hard (calcareous) plates, and that many of these bear spines. On the aboral side of the disc is a rounded body, the **madreporite**. Is it **radial** or **interradial** in position; that is, does it lie in the line of a ray or between two rays? Sketch the aboral surface, and draw a line through it dividing it into symmetrical halves. How many such lines can be drawn? The arm opposite the madreporite is known as the **anterior ray**.*

With the needle demonstrate that the calcareous plates are not on the outside. What covers them? Are the spines movable on the plates? Scattered over the aboral

---

* The reasons why this is called anterior rather than posterior cannot be worked out on the forms selected for dissection, but can only be seen by a comparison with the heart-urchins (Spatangoids), etc.

surface are numbers of fleshy, finger-like projections, the **branchiæ.** Look at the very tip of the arm, and find the rounded red **eye-spot** (recognized with difficulty in preserved material).

INTERNAL STRUCTURE.

Cut into the side of one of the arms, carrying the incision outward to near the tip, crossing to the opposite side and then back towards but not quite to the disc. Fold back the flap thus separated, and notice the following structures :

Attached to the aboral surface the lobular **hepatic cæca,** each supported by a thin membrane (**mesentery**).

On the floor (oral surface) a series of thin-walled vesicles, the **ampullæ.** By means of a needle ascertain if these ampullæ are connected with the ambulacra.

Continue the removal of the aboral surface from the rest of the body, taking care that all soft parts are separated from it and left in the oral portion, and that the portion immediately around the madreporite be left intact, and that one arm be left untouched. Now find on the aboral surface of each hepatic cæcum the **hepatic duct.** Trace these ducts inward until they enter a saccular structure, the **pyloric** part of the **stomach.** Do they unite before joining the stomach ? On the aboral surface of the pylorus is a small lobular structure, the **branchial tree.** How many branches has it ? Is it radial or interradial in position ? Draw a line through the starfish passing through the branchial tree, dividing the animal into symmetrical halves; how does this symmetry compare with that obtained from the madreporite ? Near the centre of the pylorus is the small tubular intestine (frequently torn in removing the external wall). It empties by a vent on the centre of the disc; difficult to demonstrate this in the preserved specimen. Notice the openings into the branchiæ.

Remove the hepatic cæca from one arm and find the lobular **reproductive** organs near the base of the ray. Where does this duct connect with the external wall? Would you consider this point (at which the duct opens to the exterior) as radial or interradial?

Below (that is, oral to) the pylorus is the **cardiac** portion of the stomach, produced into **gastric pouches** in each of the rays. Trace from these pouches the thin **retractor muscles** into the ray to their attachment to its floor.

Make a sketch of your dissection, showing in the centre the stomach, in one arm the hepatic cæca, in a second the reproductive organs, a third with cardiac retractors and ampullæ, a fourth with the dorsal surface, and leave the other arm for structures, to be added later.

Carefully cut away stomach a little inside the mouth, and then trace the **stone-canal** (a hard S-shaped tube) downward from the madreporite to the region around the mouth. Examine this circumoral region from the aboral side, and find the ten **Polian vesicles** (much like the ampullæ) and, inside of these, the small sacculated **racemose vesicles**. How many are there of these? What do you find in the place of the one needed to make symmetry? Beside the stone-canal is a thin-walled sac, the so-called **heart**. Sketch the organs in this paragraph, and keep the drawing for further additions.

Remove the ampullæ, membranes, etc., from the floor of one of the rays and see the **ambulacral plates** which meet in the median line. Notice the openings in this **ambulacral area** by means of which the ampullæ connect with the ambulacra. Are these **ambulacral pores** in or between the plates? How many rows of them do you find in an arm?

Sketch these plates in the ray of the drawing left incomplete.

Turn this same ray over, remove the ambulacra, and see the ambulacral plates from the oral surface. They meet, forming an **ambulacral groove**, the edges of which are formed by smaller plates (**interambulacrals**) bearing movable spines.

Cut off the arm as yet left intact about half an inch from the disc, and draw the section, including in the sketch the ambulacral plates forming the roof of the ambulacral groove; outside of these the interambulacrals, and then the plates of the aboral surface. Add to these parts the branchiæ, ambulacra, ampullæ, hepatic cæca, and mesenteries in their proper position.

In the groove of that part of the arm which remains attached to the disc notice a tube, the **radial canal**. Insert into this the canula of a hypodermic syringe or other injecting apparatus (see Appendix), and force in some colored fluid (solution of carmine or Prussian blue). What happens to the ampullæ and ambulacra ? Part the ambulacra and follow the colored radial canal to the region of the mouth, and see how this is surrounded by a **ring-canal**. Are stone-canal, racemose vesicles, or Polian vesicles filled with the fluid ? Insert the radial and ring canals, ampullæ, and ambulacra in the drawing of the stone-canal, etc.

Beneath the radial canal is a thickening of the skin, the **radial nerve** which connects with circumoral **ring-nerve** just below the ring-canal.

SEA-URCHIN: LABORATORY WORK.

What is the general shape? Are the spines movable? Can you find ambulacra between the spines? In how many areas are they arranged? At one pole of the urchin find the oral area closed by a thin membrane (**peristome**) and in its centre, teeth (how many?). Do the ambulacra radiate from this mouth? If so, where should you look for the eye-spot (compare starfish)?

In a cleaned specimen * make out the **ambulacral areas** radiating from the region of the mouth. They may be recognized by the presence of the ambulacral pores. Do these pores pass through or between the plates? How does this condition compare with that found in the starfish? Between each two sets of ambulacral plates are found the larger **interambulacrals.** Which plates, ambulacral or interambulacral, bear rounded prominences for the articulation of the spines? Making a comparison with a starfish, where would you draw the line between two rays of the sea-urchin? Illustrate by a sketch.

Follow a ray from the oral area to the pole opposite the mouth. Notice in the centre of this pole a circular **anal**

---

* For this purpose the parts of the shell (**test**) of specimens used in previous years may be employed. They are easiest cleaned by rubbing off the spines and then bleaching in Eau de Javelle or Labarraque's solution (potassium or sodium hypochlorite), to be had of druggists.

area, made up of small **anal plates.** How many plates make up the boundary of this circle? Examine them under the lens and decide which one compares in structure with the madreporite of the starfish. Is it radial or inter-radial in position? How many of these plates bear small pores? Sketch this region, showing the anal area and the tips of the rays, and label the parts, deciding which of the perforated plates must be **genital** and which must be **ocular plates** by comparing with their relative position, radial or interradial, in the starfish. With what is the madreporite associated? What parts must belong to the aboral surface of the starfish?

Internal Structure.

Open an alcoholic urchin by breaking into the equator of the test, and then continue the opening around by breaking, bit by bit, with the forceps around the shell, taking care that the fleshy parts beneath be not injured. Then carefully lift the aboral pole and, separating every-thing from it, leave all the soft portions in the oral half.

Most prominent at first will be the yellowish reproduc-tive organs occupying a position above everything else. Are its lobes connected? Can you trace the ducts of this organ? Sketch the reproductive system and then remove it. This will expose the alimentary canal (brown in color) supported by a mesentery. Trace its course, making draw-ings as you proceed. How many turns does it make? At its oral end the alimentary canal connects with a compli-cated apparatus—**Aristotle's lantern**—composed of numer-ous harder portions and muscles to move them. Have the teeth any relations to this apparatus? Look on the inside of the test for the **ampullæ,** and between them for the **radial canal.**

Usually in preserved urchins the stone-canal becomes so tender as to be easily destroyed. It goes downward from the madreporite to the inner end of Aristotle's lantern, where it connects with a ring-canal, and from this arise the radial canal, in much the same way as in starfishes, the whole forming a **water-vascular system**. As in the starfish, the nervous system follows this water-vascular system.

COMPARISONS.

With columns for starfish and sea-urchin, answer the following questions:

(1) What is the general shape of the body?

(2) Are the radial canals inside or outside the hard body-wall?

(3) Do you find branchiæ?

(4) Are all the spines movable?

(5) Is an Aristotle's lantern present?

(6) How many divisions to the reproductive organs?

(7) Are hepatic cæca present?

(8) Do you find a branchial tree?

(9) Do you find gastric pouches?

ASTEROIDA (Starfishes).

In the starfishes the flattened body is either pentagonal, or has a number of arms, or **rays** (usually five), giving it the shape of a star. In the body-wall are numerous calcareous plates, movable on one another. In the **axis** of each ray, on the side of the body with the mouth (**oral surface**), are regularly arranged **ambulacral plates,** margined on either side by **interambulacral plates** similarly arranged. In the rest of the surface (**aboral surface**) no such regularity of plates occurs. The **mouth** is in the centre of the **disc** which unites the rays, and is always without jaws or other hard parts. The mouth opens directly into a capacious stomach, the extent of which is increased by **gastric pouches.** The stomach is also partially divided by a constriction into two chambers, an oral, **cardiac,** and an aboral, **pyloric,** division. From the latter a short intestine runs to the aboral pole, where it may open by a vent, but in some no vent occurs. Into the pyloric chamber empty the ducts of five pairs of glands (**hepatic cæca**) which secrete the digestive fluids, while from the intestine arise from one to five saccular outgrowths, the **branchial trees,** the function of which is uncertain.

The organs of locomotion consist of tube-feet or **ambulacra** on the oral surface of each arm. These are connected with sacs or **ampullæ** inside the ray, and each of these systems is in turn connected by **lateral canals** with a

radial canal running below the arm in the median line.
These radial canals unite to form a ring-canal around the
mouth, and this in turn communicates with a stone-canal
which leads to the aboral surface, and thence to the exterior
through pores in a specialized plate, the madeporite. This
whole system is known as the water-vascular system. By
means of ampullar muscles the ambulacra can be extended, while ambulacral muscles serve for their retraction.
At the end of each ambulacra is a sucking-disc.

The nervous system consists, chiefly, of a nerve-ring
around the mouth and a radial nerve in each ray, the
whole paralleling the water-vascular system. Eye-spots,
one at the end of each ray, are the only specialized sense-
organs present.

The circulatory organs consist of a so-called heart beside
the stone-canal, from which vessels run in various directions,
the chief portion running between nervous and water-
vascular tracts. The only respiratory organs are the thin-
walled branchiæ, which are outpushings of the body-cavity
upon the dorsal surface.

The reproductive organs occur at the bases of the arms,
one organ on either side of each ray, the ducts emptying
in the angle between the arms. From the eggs there hatch
out larvæ which are free-swimming and bilateral, and
which show not the slightest trace of the radial shape of
the parent.

The starfishes are all marine. They feed largely on
clams, oysters, and other molluscs, and are regarded as one
of the greatest pests on oyster-beds. The way in which the
starfish feeds is interesting. It has no hard parts to break
the shell, while the mouth is too small to admit of swallowing the oyster. So it everts its stomach through the
mouth and wraps it around the shell it wishes to devour.

The retractor-muscles noticed in your dissection (p. 280) aid in pulling back the stomach after a meal.

Most of the starfishes have five rays, but this number may be exceeded, the number reaching occasionally twenty or more.

ECHINOIDA (SEA-URCHINS).

In the sea-urchins the body is spherical, heart-shaped, or disc-like, and the ambulacral areas extend, like meridians, from oral to anal regions. In short, sea-urchins are easiest compared with starfishes, if we imagine the arms of the latter bent backwards until they meet above. In this way the terminal eye-spots would be brought next to the anal area, while the reproductive openings, by the union of the arms, would be forced into a position between the oculars, and the madreporite would become pressed against one of the reproductive (genital) plates.

All of the plates are firmly united to one another, while the spines are freely movable, and share, with the ambulacra, locomotor functions. The mouth is armed with five teeth, and to aid in the movement of these a calcareous framework is found just inside the mouth, known from its first describer as **Aristotle's lantern.** In some, as in our common urchins, this framework and its muscles are complicated. From the mouth the tubular alimentary canal pursues a winding course (usually folding on itself) to the vent. Hepatic cæca, gastric pouches, and branchial trees are lacking. The reproductive organs become fused into five lobes by the union of those of the same interradius.

The Echinoida are divided into three orders:

ORDER I.—REGULARIA.

In these, which embrace the more common urchins, the mouth is at one pole, the vent at the other, and the body is approximately spherical.

288

ORDER II.—CLYPEASTROIDEA (Sand-cakes).

In the "Sand-cakes" and "Sand-dollars" we have
urchins in which the test is disc-shaped and the ambulacra
are confined to the upper surface. The mouth is in the
centre of the lower surface; the vent is on the margin

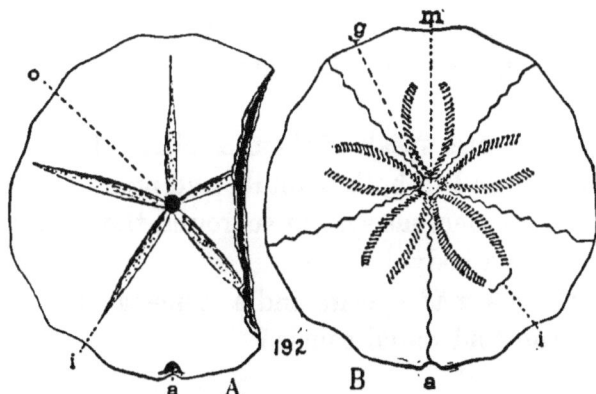

FIG. 128.—*A*, oral, and *B*, aboral surfaces of Sand-dollar (*Echinarachnius*).
a, vent; *g*, genital pores; *i*, ambulacral areas; *m*, madreporite; *o*,
mouth.

of the disc, or near the margin on the lower surface. It is
interradial in position. In a few of the sand-cakes the
margin of the disc is notched, while in others there may
be perforations extending through from upper to lower
surface.

ORDER III.—SPATANGOIDS (Heart-urchins).

In these the body, flat below, arched above, has a heart-
shaped outline, and both mouth and vent are eccentric in
position upon the lower surface. The ambulacra are all on
the upper surface, but the anterior row is lacking.

COMPARISONS.

With columns for sea-urchin and for starfish, answer the following questions:

(1) Of what is the skeleton composed ?

(2) Are spines present upon the outside of the body ?

(3) Can you speak of the parts as being radiately arranged ?

(4) Can you also speak of them as bilateral ?

(5) Do you find in both ampullæ and ambulacra ?

(6) Does the nervous system surround the mouth ?

(7) Is there a body-cavity ?

(8) Is there a madreporite and a stone-canal ?

(9) Do you find radial canals ?

This term means spiny skin, and both starfishes and sea-urchins possess this peculiarity in a high degree. But besides this external characteristic there are many other features which distinguish the group. In fact, there is scarcely a division in the whole animal kingdom more sharply marked off from other forms than this. In all the body is built on that radiate plan which is so promi-nent in starfish and urchin, and in all except a few starfish there are five rays, although in some the rays may subdivide. This radiate condition affects not only the external surface, but may extend to every sys-tem as well. And yet we may trace in every form a bilateral-

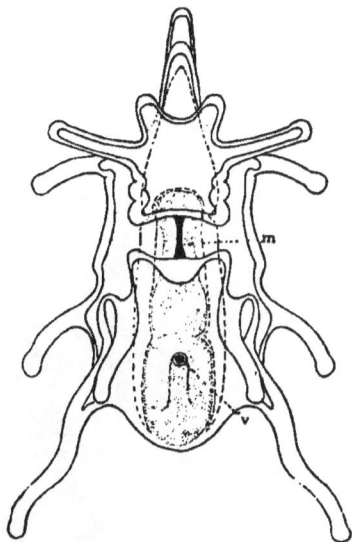

FIG. 129.—Larva of a starfish, en-larged. *m*, mouth; *v*, vent.

ity, and development shows that the bilateral condition is primitive, the radial character being acquired with growth.

Another characteristic is the ambulacral apparatus with its water-vascular system, and a third feature is the possession of a large body-cavity distinct from the alimentary canal. In all there is the formation of calcareous plates in the skin, and in all except the Holothurians these plates are developed into a more or less solid skeleton. All possess a bilateral larva, and in its development the young goes through most wonderful changes. The Echinoderms are all marine. The group is subdivisible into five classes:

CLASS I.—HOLOTHURIDEA (SEA-CUCUMBERS).

The Holothurians are cylindrical Echinoderms, with mouth and vent at the end of the body, and usually with

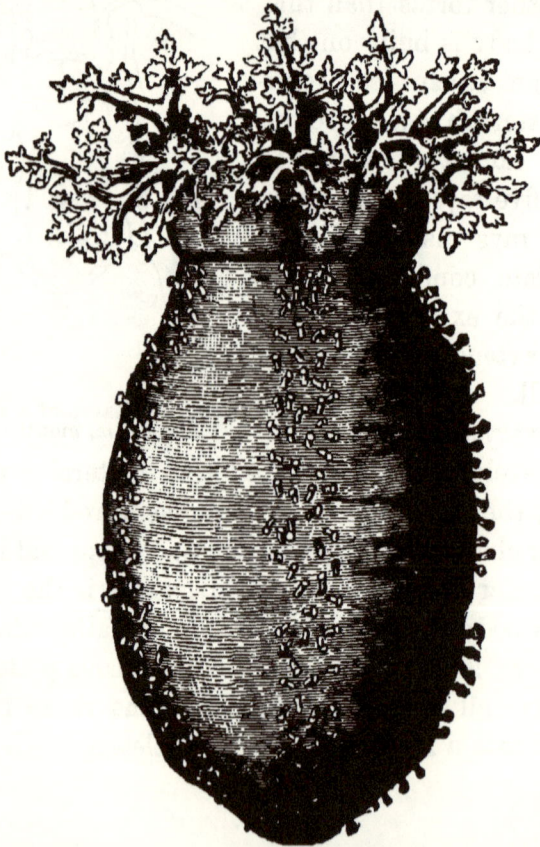

FIG. 130.—Sea-cucumber (*Pentacta frondosa*). From Emerton.

the ambulacra scattered over the surface in such a way as to make the comparison with a cucumber most apt. Around the mouth is a circle of tentacles (in reality enormously developed ambulacra), and with these the animals obtain their food. Inside, the pharynx is surrounded by calcareous plates, the whole resembling slightly the lantern of the sea-urchin, but no teeth are ever developed. In most species the madreporite is inside the body, and in many the branchial trees (p. 285) become developed into large tree-like structures. Along our shores two groups or orders occur: PEDATA, in which there are ambulacra and branchial trees; and APODA, in which both these structures are lacking, and the body is decidedly worm-like.

CLASS II.—ECHINOIDEA (SEA-URCHINS) (see p. 288).
CLASS III.—ASTEROIDEA (STARFISHES) (see p. 285).
CLASS IV.—OPHIUROIDEA (BRITTLE-STARS).

The brittle-stars, or serpent-stars as they are frequently called, are much like the true starfishes, the chief distinctions being that in the brittle-stars the arms and the disc are sharply distinct from each other, and that the extremely mobile arms are long, slender, and somewhat snake-like. A little closer examination shows that the ambulacral groove has been carried into the interior of the arms, and that here one must search for the ambulacral plates. There is no vent, and the madreporite occurs on the lower

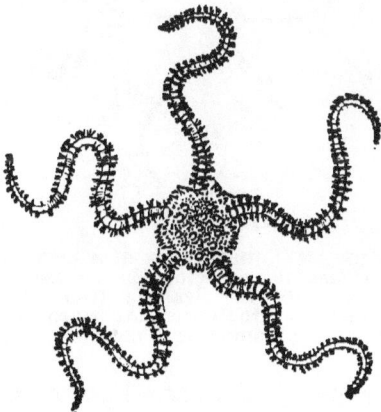

FIG. 131.—Brittle-star (*Ophiopholis*). From Morse.

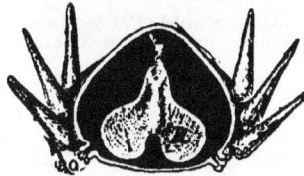

FIG. 132.—Cross-section of arm of brittle-star. *a.* ambulacral plate; *ao*, ambulacral opening.

side of the body, usually covered by one of the plates sur-

rounding the mouth.   There are a few forms in which the
arms branch again and again, and since when captured these
forms bend the arms inwards towards the mouth, giving

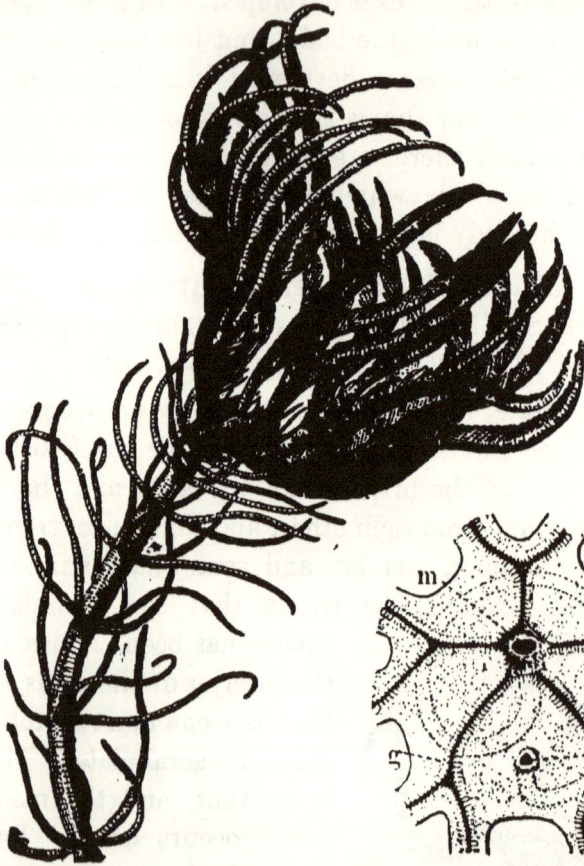

FIG. 133.—Crinoid (*Pentacrinus*) half
natural size.   From Brehm.

FIG. 134.—Mouth area of a crinoid
(*Comatula*) showing the course of
the intestine leading from the
mouth (*m*) to the vent (*a*). (*l*, grooves
leading from anus to mouth.

a somewhat basket-like appearance, these are known as
"basket-fish."   The name brittle-stars is due to the fact
that in some the arms are very easily broken.   A few
brittle-stars produce living young.

Class V.—CRINOIDEA (Sea-lilies).

While all other echinoderms are free throughout their lives, the crinoids are characterized by being fixed to some firm support by a long stalk arising from the aboral surface of the body. In most the stalk persists throughout life, but in a few, after the adult condition is reached, the body separates from the stalk and thereafter follows a free life. From the central disc or **calyx** radiate the five (usually) branching arms, and these arms and their branches bear small branchlets, so that as these animals rest in their ordinary position, the whole forms a funnel-like net with the mouth at the small end. On the upper (oral) side of all these branches run grooves converging at the mouth, and so any object which falls anywhere on the funnel is brought to the animal as food. The alimentary canal runs spirally through the calyx, and the vent is on the oral surface. The stalk, like the calyx, is strengthened by calcareous plates, those of the stalk being disc-like and piled one on another.

Crinoids, with the exception of the free forms (*Comatula*), are among the rarities of museums, as they are found only in the deeper seas. In past time, however, they were very abundant, and whole layers of rock in certain localities are made up of their remains. The fossil forms present a greater variety of shape than do the living representatives.

SEA-ANEMONE:* LABORATORY WORK.

In the prepared specimen notice that the body is cylin-drical and may be described as consisting of a column, with a **base** by which the animal was attached, and an **oral disc** bearing a large number of finger-like tentacles, in the centre of which is the **mouth**. Which tentacles, inner or **outer**, are the larger? If there be an increase in number of tentacles during growth, which ones would probably be the older? What is the shape of the mouth? How many thickened places do you find in the mouth? These thickened portions are called **siphonoglyphes**. Could they be used to indicate bilateral symmetry? Make a drawing of the animal showing the column, oral disc, etc. Cut off a few tentacles, and see if they be hollow or solid.

INTERNAL STRUCTURE.

Cut the animal with a sharp knife into two portions, the incision being made parallel to the oral disc and pass-

* It requires some patience to prepare sea-anemones for laboratory work. If merely collected and placed in the preservative fluid, the result will be a shapeless mass, in which the student will find every-thing confused. The anemones should be placed in shallow dishes of salt water, allowed to expand, and then gradually be stupefied by the addition of crystals of sulphate of soda (Glauber's salts); and then, when completely stupefied, kill and harden by transferring to a ½% solution of chromic acid for three hours. The specimens are then washed for half an hour in running water and transferred to the preservative fluid—formalin or alcohol (see Appendix).

ing through the body about half an inch from the mouth end. In the upper portion (*i.e.*, that nearest the oral disc) will be found an **œsophagus** extending inwards from the mouth. Can you trace the siphonoglyphes into this tube? Extending inwards from the outer wall to the œsophagus are six* pairs of partitions, the **primary mesenteries or septa.** The result of this is that the space inside of the body is divided into a series of chambers. The chambers between the septa of a pair are called the **intra-radial,** those between the pairs of septa the **interradial chambers.** The interradial chambers will be found to be partially subdivided by other pairs of septa (**secondary, tertiary,** etc.) which extend outwards from the body-wall, but which do not reach the œsophagus.

Examine the primary septa and find on each a **muscle** extending in the direction from oral disc to base. Are these muscles on the inter- or intra-radial sides of the septa? Examine all the primary septa before deciding this question. Then sketch the cut surface, inserting body-wall, œsophagus, and primary septa ; and on each of the septa put the muscles on the proper surface. If this be done, it will be found that there is but one plane which will divide the animal into exactly symmetrical halves. The septa through which this plane passes are the **directives.** Do they correspond in position to the siphonoglyphes? Study a few of the incomplete septa. Have these muscles like the others? At the oral ends of the septa look for openings through these partitions.

Split the other part of the animal lengthwise, and pin out under water. Notice that the œsophagus does not reach

* Occasionally variations will be found in the number and arrangement of the septa ; these exceptional forms should be compared with the more normal specimens.

the base. Could food pass from the œsophagus into the inter- and intra-radial chambers? Do you find any body-cavity distinct from the digestive cavity? Do you find any opening to compare with a vent?

Along the free edges of the mesenteries are the coiled **mesenterial filaments.** Do they present the same appearance nearer the oral disc that they do farther down? On the sides of the septa near the mesenterial filaments are the **reproductive organs** (not always plainly visible).

A HYDROID * (*Pennaria*).

Examine a colony under the hand lens or low power of the microscope, and notice the branching stem (**hydrocaulus**), bearing on their tips the fleshy **hydranths** or **zooids**. The hydrocaulus is covered with a horny, tube-like **perisarc**. Does this present any striking peculiarities? Sketch the whole colony.

In a single hydranth see that there is a balloon-shaped body, the neck of the balloon being the **proboscis**, at the end of which is the mouth. The hydranth is covered with tentacles. Is there any regularity in their arrangement? Are they all similar? Look on various hydranths for

* For this purpose it is well to have some alcoholic material, and also some mounted slides, which can be used, year after year, with successive classes. To make these mounts the alcoholic material should be washed for half an hour in water, and then stained for twenty-four hours in alum cochineal (made by soaking 7 parts of crushed cochineal insects and 7 parts of a'um in 700 of water for twenty-four hours. Then boil until the amount is reduced to 400 parts. Allow to stand twenty-four hours, filter, and add a little thymol to keep it from spoiling). After staining, the specimens should be rinsed in water and placed in 80%, 95%, and absolute alcohol for at least two hours each, and then left the same length of time in oil of clove. The best specimens may then be selected, placed upon the microscope-slides, the oil drained off, and a drop or two of Canada balsam added, and a bit of thin glass (cover-glass) placed on the specimen. The slides should be allowed to b come dry and hard (which will take some weeks) before being placed in the hands of the student. It must be borne in mind that all of the above details are necessary ; *omissions will result in failure*

globular outgrowths, **medusa-buds**. Sketch a hydranth enlarged, showing the points made out.

Study a mounted specimen under higher microscopic powers, and see that the zooids are made up of two layers and that they contain a central digestive cavity. Can you trace the layers and the cavity into the hydrocaulus? Could food taken into one of the hydranths pass to another hydranth? Are the tentacles solid or hollow? Examine the tip of a tentacle of the series nearest the mouth, and see the large oval **nettle-cells** imbedded in it. (In favorable specimens threads can be seen extending from the nettle-cells). Sketch a hydranth enlarged, showing layers, digestive tract, etc., and a medusa-bud.

Look carefully over the hydranths and see if you can find any traces of an œsophagus turned into the body as in the sea-anemone ; of septa, and of mesenterial filaments. Do any individuals show a bilateral nature?

COMPARISONS.

With columns for Sea-anemone and for Hydroid, answer the following questions:

(1) Are the animals simple or do they form colonies connected together?

(2) Can you find traces of bilaterality in the animals?

(3) Are septa present? Is the digestive cavity simple, or is it subdivided into chambers?

(4) Is there an inturned œsophagus?

(5) Are the tentacles hollow or solid?

SCYPHOZOA (Sea-anemones, Corals, etc.).

All those animals which, like the sea-anemone, have an œsophagus turned into the body (much as one might turn in the mouth of a bag), and which have the digestive cavity, subdivided by radially arranged partitions or septa, are called Scyphozoa. To these characteristic features may be added others. Thus there is a circle of (usually hollow) tentacles surrounding the mouth, and the edges of the septa bear long mesenterial filaments which are digestive in function. By the septa the digestive cavity has its surface greatly increased, so that the substances rendered soluble by the secretions of the mesenterial filaments can be more readily absorbed. Further details of structure are better given in treating of the two subclasses into which the Scyphozoa are divided; merely saying here that all are marine.

Subclass I.—Actinozoa (Sea-anemones and Corals)

In these the animal is fixed; it never swims freely, and the body in general has much the same structure as was found in the sea-anemone dissected. In some the individuals (**polyps**) are separate; in others the individuals reproduce by division or by budding, and the new polyps thus formed never completely separate from their parents, so that large aggregations or **colonies** result. In these one can distinguish the mouths, and usually the tentacles of the individual polyps, but the division does not affect the digestive tract, so all are connected, and the food which is taken in at one mouth may serve to nourish any part of the whole colony. In some the outer surface of the body is naked, but in many of the solitary and in most of the colonial forms the base or both base and column secrete carbonate of lime, thus forming a solid support for the body. This solid support is the well-known **coral**. In most spec-

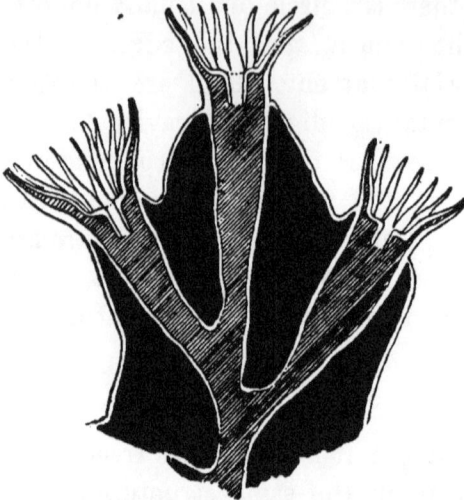

FIG. 135.—Diagram of a bit of coral to show the way in which the polyps are connected. The coral is black ; the digestive cavity shaded.

FIG. 136. — Section of a coral cup showing the calcareous septa. After Pourtales.

imens of coral one can readily recognize the cups in which the separate polyps were situated; and in these cups, in most cases, are calcareous partitions much like the septa of the soft parts.* As long as the colony remains alive it is constantly budding off new polyps, and thus the colony and the coral grow. Those species which live in cold water produce but little coral, but in tropical waters the coral-producing forms abound, and by their combined secretions the coral islands are made.

The great majority of the Actinozoa may be subdivided, according to the number of septa, into two orders:

* These calcareous septa do not coincide with, but alternate in position with, the fleshy septa.

Order I.—Octocoralla.

In these the separate polyps are small, and each has but eight septa and eight tentacles. They produce but little coral, but rather those kinds of coral which are known as sea-fans and sea-whips. One form is especially noticeable since it produces the precious red coral so often carved into beads, etc.

Order II.—Hexacoralla.

As the name indicates, the septa and tentacles here occur in multiples of six. Here belong all the sea-anemones and the true corals which produce coral-reefs and islands. The

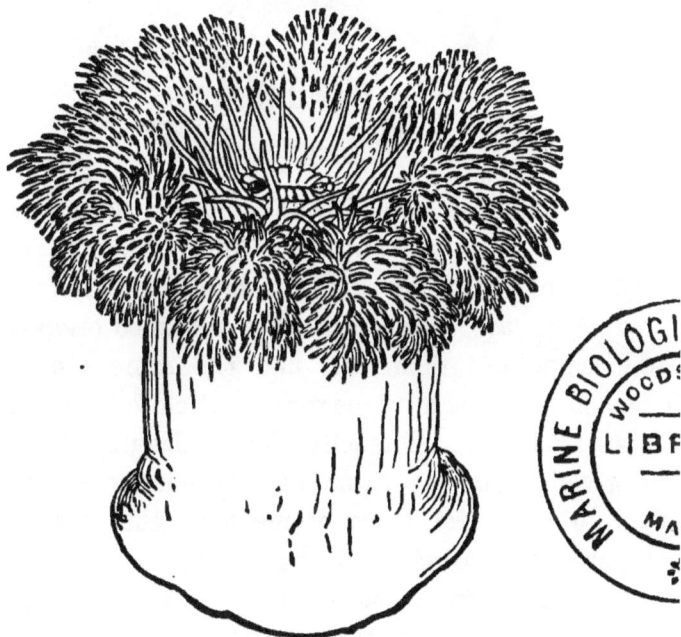

Fig. 137.—Sea-anemone (*Metridium*). From Emerton.

reef-building species are limited in their distribution by temperature, for they cannot live where the temperature of the water falls below 60° Fahr. (13° C.).

SUBCLASS II.—SCYPHOMEDUSÆ (Jellyfishes).

At first sight these Scyphomedusæ differ greatly from the Actinozoa. They are free-swimming forms in which the body is developed into an umbrella-shaped structure,

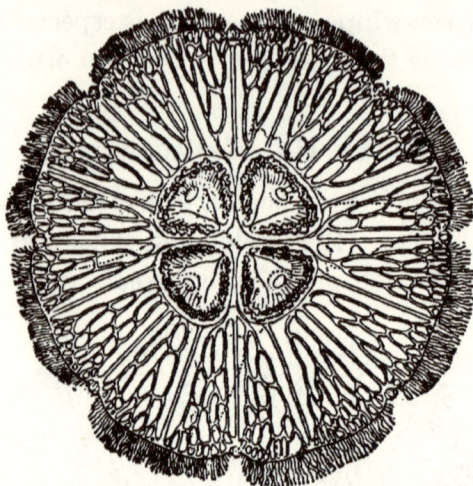

FIG. 138.—Common white jellyfish (*Aurelia*). After Agassiz.

the mouth is at the end of a long proboscis, and all is semitransparent. Yet when the details of structure are analyzed there are found the same inturned œsophagus, the same septa and filaments, and the same tentacles; and hence these forms must be somewhat closely associated with the sea-anemones. The jelly-like consistency of their bodies has given them the name of jellyfishes. A rather more accurate name for them is **medusæ**, the tentacles being compared to the snaky locks of that mythical monster. They swim through the water by a lazy flapping of their umbrellas, feeding upon whatever may come in their way. While some are small, others become veritable giants, the large blue jellyfish of the New England coast sometimes measuring seven feet across; its tentacles streaming behind for a hundred feet as it swims through the water.

HYDROZOA (HYDROIDS, ETC.).

In contrast with the Scyphozoa the Hydrozoa lack the inturned œsophagus and the septa dividing the digestive tract, while they have solid tentacles. In size they are on the average much smaller, and colonial forms predominate. In their life-history we frequently meet some wonderful changes, and to describe these we may follow through the life-cycle of the Pennaria, studied in the laboratory work.

From the egg there hatches out a little oval, free-swimming embryo, which soon attaches itself by one end to some submerged rock, while a mouth breaks through at the other, and tentacles grow out around the sides of the body. When a mouth is formed feeding and growth are possible. As the animal grows larger little buds appear on the sides, and these, forming mouths and tentacles, grow into hydranths like the parent. These buds never become free, but the whole colony thus formed has a common digestive tube by which all are connected. On the outside a tubular protecting sheath, the perisarc (p. 299), is developed. After a while buds appear on the sides of the hydranths, and these have a much different history, for they develop into free-swimming jellyfishes.

These jellyfishes (see Fig. 139) are much like those of the Scyphomedusæ (p. 304), being bell-shaped or umbrella-shaped, the mouth being at the end of the handle, while branches of the digestive tract run to the margin of the umbrella. However, these hydrozoan jellyfishes differ from the Scyphomedusæ in the absence of mesenterial fila-

305

ments and inturned œsophagus, and are further distinguished by the fact that the aperture of the umbrella is
partially closed by a thin membrane (**velum**) lacking in the
other jellyfishes.   These jellyfishes produce the eggs from
which are developed other colonies like that studied.

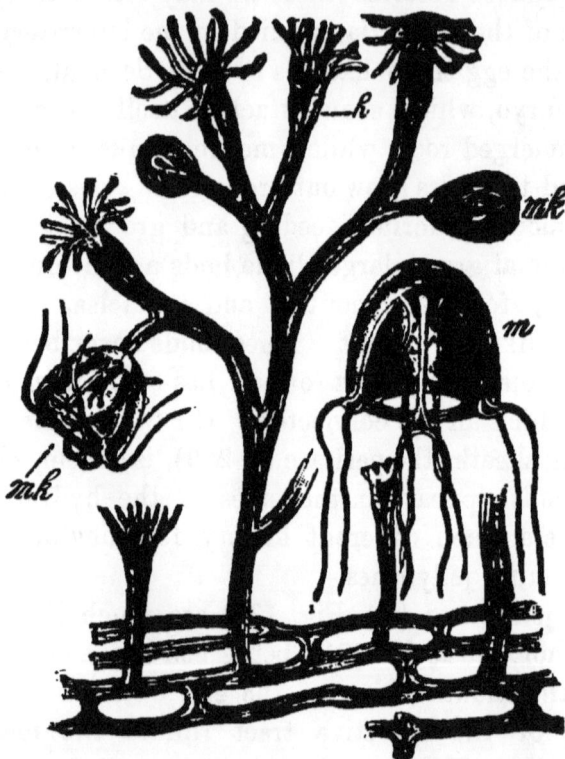

FIG. 139.—A Hydroid (*Bougainvillea*. After Allman, from Lang. *h*, feeding polyp; *mk*, medusa-buds; *m*, a free-swiming medusa. At the base
are seen the root-like stolons connecting the colony together.

Here is a point which needs emphasis.   From the egg is
developed a hydranth which by budding develops numerous other hydranths, and each of these in turn, by budding
produces several medusæ.   In other words, we have here
an animal which reproduces asexually.   These medusæ are

the sexual forms, and they produce eggs which grow not into other jellyfishes but into the fixed forms. This phenomenon is known as an **alternation** of **generations**, the young resembling not the parent, but, rather, the grandparent.

ORDER I.—HYDRIDÆ.

Here belongs the fresh-water Hydrozoan—the Hydra—in which there is no medusa stage, the animals producing eggs which develop directly into other Hydræ. The fresh-water Hydræ, which are green or brown in color, and about a quarter of an inch in length, abound in fresh water

ORDER II.—HYDROMEDUSÆ.

The Pennaria is typical of this group. In most there is that alternation of fixed and free-swimming forms which has already been described. In the fixed stage the colony is usually protected by a perisarc which occasionally may be developed into cups protecting the hydranths. On the other hand some of these Hydromedusæ exist only as jellyfishes, the eggs which they produce developing directly into other jellyfishes. The Hydromedusæ are abundant in all seas, and are among the most beautiful and interesting of all the animals with which the naturalist has to deal. Only two or three species occur in fresh water.

ORDER III.—SIPHONOPHORA.

These may be defined as colonies of jellyfishes, arising by budding. In these colonies the medusæ become specialized in different directions. This specialization in some forms may go so far that we have the jellyfishes modified into (1) a float supporting the colony; (2) swimming-bells by means of which it moves; others (3) for feeding, still others (4) for digestion, and again others (5) for re-

production, etc.   Usually one or more of these is absent
from the colony.   The most familiar of the Siphonophores

FIG. 140.—Diagram of a Siphonophore.
  c, covering scale; d, digestive sac;
  f, float; m, mouth of feeding individual; r, reproductive bell; s,
  swimming-bell; t, tentacle.

FIG. 141.—Portuguese man-of-war
  (*Physalia*). After Agassiz.

is the "Portuguese man-of-war," which occasionally drifts
on our shores.   In this beautifully-colored species the float
is large and the swimming-bells are absent.

COMPARISONS.

Answer the following questions for sea-anemone and for hydroid:

(1) Has either a radial arrangement of parts ?

(2) Are tentacles present ?

(3) Is there a body-cavity apart from the digestive cavity ?

(4) How many openings to the digestive tract ?

The Cœlenterata and the Echinoderma were formerly united into a group Radiata, the basis of association being the radiate type of structure so noticeable in a starfish or a coral. Later studies showed that the two divisions had very few points in common, and that the differences between them were very great. Remembering the account of the Echinoderms, the following facts regarding the Cœlenterata will have significance:

In the Cœlenterates there is but a single opening into the digestive tract, which thus serves at once for mouth and vent. Through it all food enters, and all indigestible portions are cast out. The mouth connects with the digestive tract, which extends to all parts of the body, so that the food is brought close to every portion, there being no circulatory apparatus. There is no body-cavity distinct from the digestive tract. Around the body, usually close to the mouth, is a circle of tentacles, and on these abound some structures which need a slight description —the nettle-cells.

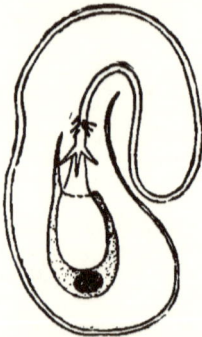

FIG. 142. — A discharged nettle-cell, the thread coiled around the cell.

These nettle-cells are small bodies which occur all over the body, but are especially numerous upon the tentacles. Each is in reality a sac, one end of which is drawn out into a long and slender tube, coiled up inside of the rest. These nettle-cells can be "discharged" by the animal, and this discharge consists in a forcing out of the tube in the same way in which one may blow out the inturned finger of a glove. These cells contain a strongly irritant poison, and at the discharge this poison escapes. These nettle-cells furnish a means of defense, and they are also used in obtaining food, the poison being strong enough to paralyze instantly small animals. In some forms it is strong enough to affect man. For instance, the tentacles of the Portuguese man-of-war will quickly raise a bright red ridge on the hand or arm of man, and produce an almost intolerable burning sensation in the parts thus touched.

In many Cœlenterates there is no specialized nervous system, the general surface of the body having sensory and nervous powers. In others there is a central nervous system arranged in a ring around the body; and some of the jelly-fishes may have organs, the structure of which leads to their being regarded as simple types of eyes and ears.

With very few exceptions, the Cœlenterates are marine. Some move about freely, some are as firmly fixed as is any plant; but, as has already been explained (p. 305), both fixed and free conditions may occur in the life-history of a single species. All of the Cœlenterates reproduce by means of eggs, but, besides, most forms have the power of forming buds which grow into new individuals, sometimes like, sometimes greatly different from, the parent. The Cœlenterates are divided into two classes, accordingly as they possess or lack an inturned œsophagus.

CLASS I.—HYDROZOA (p. 305).

CLASS II.—SCYPHOZOA (p. 301).

Usually a number of free-swimming jellyfishes are closely

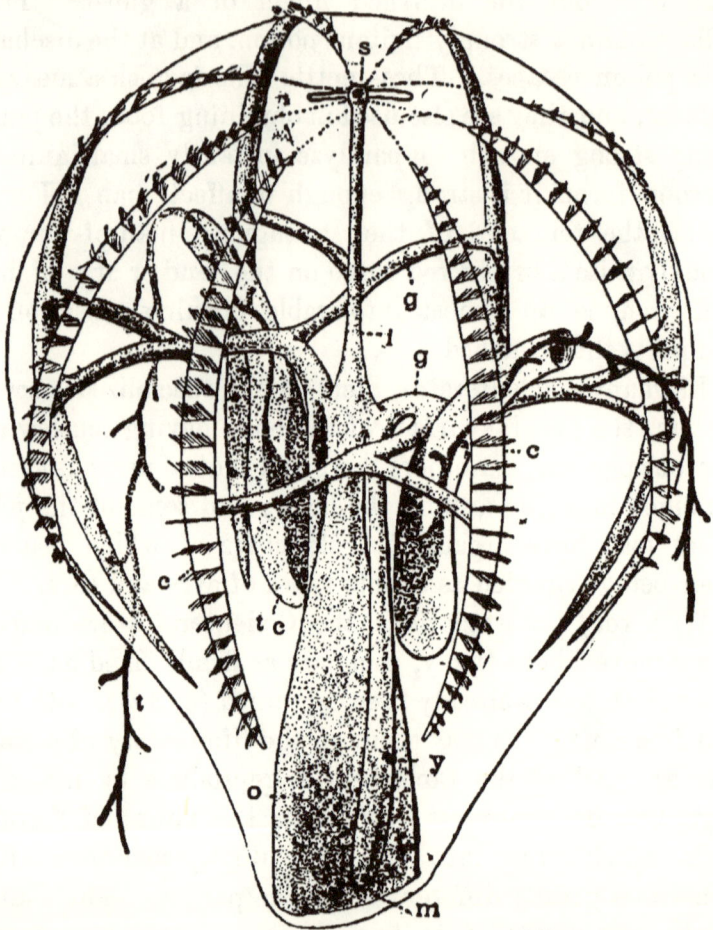

FIG. 143.—Diagram of a Ctenophore. *c*, rows of combs; *g*, branches of digestive tract; *i*, "intestine"; *m*, mouth; *o*, œsophagus; *s*, sensory plate; *t*, tentacle; *tc*, tentacle sheath; *y*, œsophageal vessel.

associated with the Hydrozoa and Scyphozoa under the

name Ctenophora. They receive this name (meaning comb-bearers) from the fact that the usually globular body has eight rows of vibratile organs, each row being composed of series of hairs arranged much like the teeth of a comb. They have an alimentary canal, which branches so that a portion underlies each row of combs. They, however, differ from all true Cœlenterates in the absence of nettle-cells, and in a number of other features which need not be described here. All of the Ctenophores are marine.

SPONGES : LABORATORY WORK.

A.—A Calcareous Sponge (*Grantia*).

Notice the shape. Is the surface smooth? How many openings do you find? What differences do you find between the ends? Split the sponge lengthwise with a sharp scalpel, laying open the central cavity (**cloaca**). Where is the large opening (**ostium**) by means of which the cloaca is connected with the exterior? By what is it surrounded? In the walls of the cloaca notice the openings (**excurrent canals**)—best seen after the sponge has dried. In the cut walls see the small chambers (**ampullæ**). Draw one half of the sponge, naming the parts. Cut the other half of the sponge transversely, and notice the radially arranged ampullæ. Place a bit of the sponge in weak hydrochloric acid. What occurs? Boil another bit in caustic potash (a few drops of a 5-per-cent solution), then place the fluid on a slide; examine under the microscope. Draw the **spicules** which you see. Crush a dry bit of the sponge in the fingers. Has it any elasticity?

B.—A Bath Sponge.

Select small rounded sponges for this purpose. Notice the irregularity of the surface. Do you find any large opening in any way comparable to the ostium of the calcareous sponge? If so, split the sponge through this opening and study the section. Can you find canals branching from the ostium? If so, sketch their **arrangement**.

314

Crush a bit of the dry sponge between the fingers.   How does it compare with the other form ?   Examine a very thin bit of it under the microscope; can you find spicules ?

It is to be noted that in the bath-sponge only the hard or skeletal parts are present, the flesh having been washed away.   In the calcareous sponge, as put up for laboratory use, flesh and skeleton are both present.

SPONGES (PORIFERA).

Sponges differ from other animals in so many respects that for a long time naturalists were uncertain as to whether they were animals or plants, but this matter has long been settled beyond dispute.

All sponges are composed of the following parts: On the outside are numbers of small openings or pores (whence the name Porifera), and these lead to small tubes or **incurrent canals**, which extend inward to small chambers (**ampullæ**), the digestive organs, which also contain the apparatus for keeping up the flow of water through the canals. With the water numerous small particles of food are drawn into the ampullæ, and are there taken up, while the water leaves the chambers by means of a second system of tubes (**excurrent canals**), passes into a central space (**cloaca**), and thence to the exterior by a large opening (**ostium**). This same system of canals also serves for respiration, but special muscular, circulatory, nervous, sensory, and excretory organs are lacking.

In a few sponges there is no skeleton, but most species have a firm support for the soft parts. This skeleton may consist of small particles (**spicules**) of carbonate of lime or of silica, often much like crystals in form; or of fibres of a horny substance; or again, both spicules and fibres may occur together. In the sponges of the stores we have nothing but the horny fibres, all of the flesh having been washed away; but in this skeleton we can trace roughly the systems of canals, the cloaca, and the ostium.

Sponges reproduce by budding and by eggs. In budding

smail outgrowths occur, and these gradually become larger, and finally an ostium is formed. From the eggs are formed little free-swimming embryos, which later settle down and grow into the adult.

Sponges are largely marine, only a few forms, and these of no economic importance, occurring in fresh water. The sponges of commerce come from the Mediterranean, the

FIG. 144.—Sponge (*Dactyocalyx*). From Lütken.

Red Sea, and Florida and the West Indies. They are brought up by divers, or by hooks which are dragged over the bottom. The fleshy portions are allowed to decay, then the skeleton is washed, and the sponges are packed in bundles for the market. There are different grades of elasticity and fineness of fibre, and consequently different values. The finest sponges come from the eastern part of the Mediterranean. Sponges occur as fossils, especially in the Cretaceous rocks.

There are two great groups of sponges. In the first, called CALCAREA, the skeleton is composed of carbonate of lime; in the second, SILICEA, there is sometimes a skeleton consisting of silica (quartz), sometimes of horny fibres, sometimes of both horny fibres and siliceous spicules; and again, there are a few forms which have no skeleton.

All of the divisions or groups of animals so far studied are united by naturalists under the name Metazoa for the following reasons: A careful consideration of their structure leads to the conclusion that, in all, the body is of appreciable size, and that, in each and every one, certain portions or **organs** are specialized for the performance of certain functions necessary in the economy of the individual. Thus we find in all reproductive organs which have solely to do with the perpetuation of the species; in all (except a few degenerate parasites) there is a mouth for taking in of food and an alimentary tract for its digestion; in all there is a more or less distinct nervous system; and in all, parts of the body are more or less specialized for respiration.

A little deeper insight leads to another conclusion which farther justifies the group of Metazoa. In all animals so far studied the body is composed of layers, at least two in number; one on the outside forming the skin, and a a second on the inside forming the lining of the digestive tract. To these two layers are given names, **ectoderm** and **entoderm**, meaning respectively outer and inner skin.

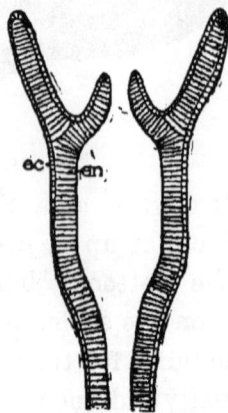

FIG. 145.—Diagram of a two-layered animal, based upon a hydroid. *ec*, ectoderm; *en*, entoderm.

In the Cœlenterata all of the functions of the animal are performed by either one or the other of these two layers. In all the other divisions a third layer occurs between ectoderm and entoderm — the **mesoderm** (middle skin), and this mesoderm takes some of the functions which are divided between

318

the ectoderm and entoderm of the Cœlenterata. The study of the development of these *three-layered* animals shows a very interesting fact. At first there are but two layers in the body, and later the mesoderm develops between these two. In other words, all of the higher Metazoa pass through a stage in which they exhibit a cœlenterate condition.

These three layers reach their highest condition in the Vertebrates, and it may be interesting to see how all the various structures which have been studied in a shark or in a rat are related to these layers.

To the ECTODERM belong the outer layer of the skin, the outer layer of scales, the hair, feathers, sweat-glands, the enamel of the teeth, the nervous system, the sensory portions of sensory organs, and the lens of the eye.

The ENTODERM furnishes the lining of the alimentary canal, the notochord, gills, tracheal lining, lungs, liver, pancreas, urinary bladder.

The contributions of the MESODERM to the body are more extensive. They include the deeper layers of the skin, fat, muscles, connective-tissue, cartilage, bones, ligaments, blood-vessels, blood, the lining (pleural and peritoneal membranes) of the body-cavity, the deeper layer of the scales, the dentine of the teeth, the outer layers of the alimentary canal, and the reproductive and excretory organs and their ducts.

If we study any part of any one of the animals already dissected or mentioned under the higher powers of the microscope—having first treated it so as to bring out details —we will discover another fact of great importance. Every one of these animals will be found to be made up of small parts, essentially like each other, just as the wall of a building is built up of separate bricks. These separate parts or elements of the Metazoa are known as **cells**. Each one of these cells is microscopic in size, with an average diameter of about $\frac{1}{2500}$ of an inch; and each consists of a semi-fluid substance known as **protoplasm**, in the centre of which is a mass of slightly different protoplasm, known as the **nucleus**. Now, since each and every metazoan is built up of cells, we may speak of the Metazoa as *many-celled animals*.

These cells vary greatly in shape, but no matter how different they may appear at first sight, they all agree with the description given in the last paragraph. Some may be spherical, others cubical or flattened, and still others branched, yet in all there is the same nucleus. Cells of the same general shape are united together to form **tissues**, so that we have bone-tissue made up of what may be called bone-cells; muscular tissue, of muscle-cells; and nervous tissue, of nerve-cells; etc.

In the Metazoa the tissues are built up into **organs** for the performance of certain purposes; and usually a single organ is composed of several kinds of tissues, while the same kind of tissue may reappear in different organs. Thus the hand of man is an organ of grasping; in it we find muscular, bony, connective, and nervous tissues; while in the heart of the shark muscular, connective, and nervous tissues appear.

The Metazoa are subdivided into groups or "branches" which may be arranged in order of their complexity in the following manner:

BRANCH   I.—CŒLENTERATA (p. 310).
BRANCH   II.—SPONGIDA (p. 316).
BRANCH III.—VERMES (p. 242).
BRANCH   IV.—MOLLUSCA (p. 269).
BRANCH   V.—ARTHROPODA (p. 226).
BRANCH   VI.—ECHINODERMA (p. 291).
BRANCH VII.—CHORDATA (p. 153).

In contrast to the Metazoa, which have just been defined as animals made up of many cells, and these cells arranged in two or three layers and grouped into tissues, comes the group of Protozoa, which may be defined as animals each consisting of a single cell. A little thought will show that this difference is in reality very great. In the metazoan certain groups of cells become adapted (specialized) for the performance of certain work in the body, and the more specialized they become the more restricted are they in their lines of work. Thus in man the cartilage and bone-cells are solely for the support of the body, muscle-cells for the moving of parts or of the body as a whole. When, however, we turn to the Protozoa, composed of but a single cell, we find that this one cell has to do all the work which in the Metazoa is shared by the several groups of cells. It has to feed, to move, to excrete waste matters, and to reproduce its kind. In a word, the cells of the Metazoa are **differentiated** in various directions; those of the Protozoa are undifferentiated.

The Protozoa show great variety in shape, appearance, and habits. In some there is no differentiation between the different regions of the cell which composes the body, excepting the fact that a nucleus is usually (if not always) present. Food may be taken in at any point; any portion may be used for locomotion; and indigestible portions may pass out anywhere on the surface. By feeding they grow,

321

and when growth reaches a certain limit the animal (cell) divides, and we have now two individuals in the place of the original one.

In other Protozoa different regions in the cell may be specialized in different directions. A single example must suffice. In the form figured we have but a single cell, but it is a cell of definite shape. Externally the body is covered with a denser layer, comparable in position and use to a skin. A little deeper are developed longitudinal lines of contractile material which act in the same way as the muscles of the Metazoa, moving one part on another. Over the outer surface are minute hair-like organs (**cilia**) which are in constant motion, and when the animal casts itself loose these serve like so many oars to propel it through the water. At the larger end of the body these hair-like organs become much larger, and they are here arranged in a spiral. The effect of their constant motion is to create a minute whirlpool in the water, the centre of which is in an opening in the larger end. This may be compared to a mouth. The water brings with it minute particles suitable for food, and these pass through the mouth into a cavity comparable to a gullet, from which they pass into the central part of the cell, where they are digested. Then the indigestible portions are at last passed

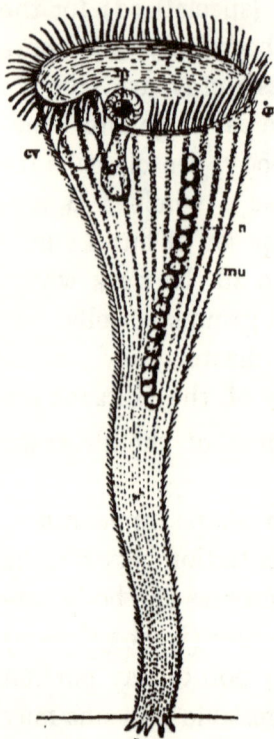

FIG. 146.—Diagram of a Protozoan based upon Stentor. *c,* large cilia around the oral disc; *cv,* contractile vacuole; *g,* gullet; *m,* mouth; *mu,* muscular bands; *n,* nucleus; *nr,* nerve-ring.

out from the body at a fixed point, the functional vent. The large cilia always move in a regular and rhythmic manner—a fact which would imply that they were connected and controlled in some manner in their action; and high microscopic powers shows at their bases a cord of somewhat denser material which takes the place of a central nervous system. If this be cut, the cilia no longer work in harmony. Finally, all animals, in doing work, produce nitrogenous waste, which must be gotten rid of by means of kidneys. In the form figured the kidneys are replaced by a space on the interior (**contractile vacuole**) which regularly enlarges and contracts, and at each contraction this waste is forced out into the surrounding water. All of this is in a single cell.

The Protozoa, of which many thousand different kinds have been described, are all very minute, only a very few being even visible without a microscope. The great majority are aquatic, some being found in fresh water and many in the sea. A few live in moist earth, and more are parasitic in other and higher animals, where they may at times be productive of disease.

COMPARATIVE PHYSIOLOGY.

An animal is a machine, and the preceding laboratory exercises are intended to give a student a knowledge of the different kinds of mechanism in the several animal types. Our knowledge of a machine is not complete when we know its structure; we must also understand the way the different parts perform their work. The study of the structure of an animal is the province of anatomy, while that branch of science which deals with the action of the various parts —the working of the whole—is called physiology.

It is a far more difficult task to ascertain from the specimens themselves the function of the parts and the action of the animal machine as a whole, than it is to make out the details of structure and so a general summary is given here.

Any and every machine, in order that it may perform work, must be supplied with energy, and the animal obtains this energy by the slow combustion (oxidation) of food, just as the steam-engine gets its energy from the rapid combustion of coal. In the case of a steam-engine there is an engineer who supplies the fuel, regulates the action of the parts, and disposes of the waste. The animal must be its own engineer. It must have the means of obtaining fuel (food), of putting it in such position that all the energy produced by its oxidation can be utilized to its fullest extent, and all waste can be properly disposed of. This has led, in the first place, to the formation of a digestive tract, in which the food is put in such shape as to be most advantageously used by the organism.

324

In the lowest animals (lowest Protozoa) we find that the whole body (cell) serves as a digestive tract, and that food can be taken in at any point on the surface. A little higher (p. 322) an organ which we must call a mouth is formed in the body, and this opening for the taking in (ingestion) of food is found in all higher animals, except a few parasites which, living in liquid food, need no such opening. With larger animals a definite digestive cavity or canal is formed, the lining of which has certain definite work to perform. Most articles of food are insoluble as taken into the body; a bit of meat or starch can be soaked indefinitely in pure water or can even be boiled for days without passing into solution. In the digestive tract juices are produced which alter these substances so that they can be dissolved; and it is only when they are in solution that they can pass through the walls of the alimentary canal to those parts where they are to be utilized.

In the lower animals all parts of the digestive tract seem able to act at once as formers of digestive fluids, and in the taking up of the dissolved food, but as we pass higher in the scale complications of various kinds are introduced. In the first place, we find certain organs, like the salivary glands, stomach, pancreas, and liver, set apart for the secretion of digestive fluids, and even in animals as low as the sea-anemone the mesenterial filaments appear to have the same power. On the other hand, the other portions, while they may secrete, are pre-eminently the region for the absorption of the liquefied food. Another complication is this: A given amount of surface can absorb only so much in a given time; so to obtain the necessary amount of food the surface must be increased. This explains in part the folding of the wall of the digestive tract in the sea-anemone, as well as the lengthening and coiling of the intestine in tadpole and rat, and the spiral valve in the shark. In many vertebrates the surface is still further increased by numerous minute foldings and outpushings of the lining of the intestine which, though so small as to be invisible to the naked eye, still more than double the surface.

With most food there are certain portions which are indigestible. These of course must be gotten rid of. In the Cœlenterates and flatworms the only opening through which they can pass out is the same one by which they entered, and so this opening, usually called the mouth, serves at once as mouth and vent. In the higher forms the alimentary canal becomes a complete tube with two distinct openings, one—the mouth—for the taking in of food, the other—anus or vent—for the ejection of non-nutritious portions.

After its solution the food (nourishment) must be transferred to the parts which are to do the work. In the Protozoa the same parts which digest do the work. In the sea-anemone and flatworms the pouching of the digestive tract renders the transfer easy, for the pouches extend to all parts. Above these forms we find circulatory organs present, one of the functions of which is the carrying of the dissolved food from the digestive tract to the working parts. These circulatory organs are tubes through which the fluid flows, but a flow can only be produced by some mechanism which shall propel the fluid. In most cases this is effected by muscles in the walls of the vessels, which by waves of contraction force the fluid along. The tendency is constantly towards a concentration of these pumping muscles in one region, and thus a heart results.

So far, we have traced the fuel to the working parts. In order to do work the fuel must be oxidized, and this means that oxygen must also be brought to these parts.

This oxygen is found either in the air or dissolved in the water in which the animal lives. In the Cœlenterates, flat-worms, and many other forms, the general surface of the body is sufficient for the absorption of the oxygen, but where the animal is larger and needs more oxygen special provisions are needed.

A very simple condition, physiologically, is found in the insects, where air-tubes (tracheæ) extend inwards from the outside, their fine branches reaching to every part of the body. Air is drawn into these tubes by an enlargement of the body by suitable muscles, and then, when the oxygen is absorbed, contraction forces out the remainder. This breathing process can be seen by watching the abdomen of a grasshopper or a wasp. In many crustacea, molluscs, worms, vertebrates, the conditions are more complicated. In these the nutrient fluid is also the bearer of the oxygen; and, in order that the fluid may obtain this element specialized portions are developed, where the circulatory fluid may come into close relationship with the water (gills) or the air (lungs). In some (see the figure of Doris, p. 276) the gills project freely into the water, and there is no special apparatus for changing the fluid. In other forms the gills are protected by enclosure in a branchial chamber, and then the water containing the oxygen must be brought here. In the oyster and clam this is effected by numerous minute hair-like structures (cilia) which by their constant motion draw water over the gills. The squid gets its supply by enlarging and contracting its mantle-cavity, the crayfish by pumping water over the gills by means of its "gill-bailer," and the fish and tadpole by taking water into the mouth and forcing it out through the gill-slits. The lungs of the higher vertebrates possess a resemblance to the tracheæ of the insects in that air is drawn into them; but here the similarity ceases, for in the vertebrates the air is brought from the lungs to the working parts by the intervention of the nutrient fluid (blood).

The methods by which air is drawn into the lungs vary. The frog swallows the air by aid of the muscles extending across the throat between the halves of the lower jaw, and that this swallowing is the only way of forcing air into the lungs is shown by the fact that if the mouth be kept from closing the animal will suffocate.* In the Sauropsida the muscles between the ribs and those forming the walls of the abdomen are concerned in the inspiration and expiration of air; while in mammals the muscular partition (diaphragm) which divides the body-cavity becomes an efficient organ in the process.

We naturally think of work in terms of motion, and in the case of an animal the contraction of a muscle or the movement of a part or the whole of the body naturally suggest themselves as examples. These, however, are but a part of the work which the animal does. The performance of any function of the body is really work. When a gland secretes, a nerve acts, an intestine absorbs, or the mind carries on its operations, the expenditure of energy is called for just as in the contraction of a muscle. So all parts must have both food and oxygen.

When coal is burned in an engine, besides energy there is a production of waste. A part of this waste passes off in a gaseous condition as water vapor and part as ashes. When any part of the animal body works there is a similar formation of waste, and the carbon dioxide and water vapor are carried away by the same structures (tracheæ in the insects, blood-vessels and gills or lungs in many other forms) which brought the oxygen to the parts.

* The skin is a very important organ in the respiration of the Batrachia (see p. 50).

The animal, besides, needs for its fuel substances known to the chemist as nitrogenous food, and the combustion of this produces, besides the carbon dioxide and water, nitrogenous waste, and this, in all of the higher animals, is gotten rid of by means of organs which can be grouped under the common name of kidneys. Here are to be placed not only those structures specifically called kidneys in the foregoing pages, but also the green gland of the crayfish, the Malpighian tubes of insects, the nephridia of the earthworm, and the organ of Bojanus in the clam. Even the contractile vacuole of the Protozoa is to be regarded as an organ for the excretion of nitrogenous waste.

We have seen that the fluid propelled by the heart has a large series of different purposes to fulfil. It must carry nourishment from the digestive tract to the different parts of the body; it has to carry oxygen from the gills and lungs to these various structures, and to carry the carbon dioxide and water produced by work to the same lungs and gills, while the nitrogenous waste must be carried to the kidneys. The fluid which does all this is the blood.

There is another aspect of animal physiology to be reviewed. The animal needs to be aware of the presence of food and of the proximity of things injurious to it. This implies the formation of a sensory system, and naturally this system must be on the outside of the body, for from without comes both food and danger. The knowledge of the presence of good or of evil would be of little value to the animal were it without ability to avail itself of this knowledge. Hence this sensory system is connected with a nervous system; which directs and controls the actions of the animal. In the lower animals this nervous system is on the surface, but as this superficial position is dangerous to such an important structure, we find in all the higher animals that it becomes removed to a deeper position, which necessitates the development of nerve-cords to connect it with the sensory system and with the muscles and other parts. It is interesting that in all animals, even in man, no matter how deeply situated or how thoroughly protected it may be in the adult, the central nervous system arises from the outer surface, and secondarily attains its permanent position.

Since most animals must search for their food, we find that one end becomes adapted for always going in front, and in this way a head has come into existence, and here are situated the brain and the most important sensory organs, as well as the mouth, since this part of the body first comes into the neighborhood of substances useful as food or likely to be injurious to the animal.

So far we have treated of the animal as an automatic self-regulating machine, but in one respect it differs from all machines of human production. No amount of fuel put under the boiler of a steam-engine will cause this mechanism to increase in size or to give rise to other bits of mechanism like itself. The animal machine grows by the taking in of food, and like the steam-engine, it wears out. It, however, has the power of reproducing the kind, by the formation of small parts (either buds or eggs) which eventually grow into animals like the parent which produced them, and thus the species is perpetuated, the young taking the place of the generation which has worn itself out.

MORPHOLOGY OF ANIMALS.

We are now in position to review some of the facts we nave already discovered, and to draw some general conclusions.

Excepting some Protozoa, each and every animal can be placed under one of two heads. In the one, the body is bilaterally symmetrical. In it we can recognize anterior and posterior; dorsal and ventral; right and left. Under the other we place those forms in which these features do not exist; there is no right and left, but the parts are radially arranged around an axis, like the spokes around the axle of a wheel. To this latter group belong the sponges and cœlenterates; to the first, all other divisions reviewed in this volume. Even the Echinoderms belong to the bilateral type, for their development shows that in the early stages they have not a trace of radial symmetry, but only acquire it later in life.

In the bilateral animals, in turn, two types can be recognized: the segmented and the unsegmented. The segmented forms show their peculiarities in the most striking manner in some of the Annelids, like the earthworm (p. 235). In these the body is made up of a series of rings or segments, each essentially like its fellow, and each containing a portion of all systems of organs—muscular, nervous, circulatory, digestive, excretory, etc. In the arthropods this segmentation again appears, but here there are tendencies in two directions: towards a fusion of segments, and towards an increase of one segment at the expense of another. In annelids and arthropods this

331

segmentation is visible externally; in the vertebrates it is not so plainly shown, but it nevertheless exists. The trunk muscles (see p. 11) are thus arranged; the spinal nerves and the vertebræ correspond to the muscle segments, as do also certain blood vessels (intercostals), while in their early history the kidneys are segmentally arranged.

On the other hand, the lower worms show no traces of segmentation, while the molluscs show it to a very slight extent.* In the echinoderms there is a repetition of ambulacra and ambulacral plates, but this is supposed to be different in its origin from that in the segmented animals.

All animals above the Protozoa reproduce by eggs. These eggs, when carefully studied, are found to agree in their essential characteristics. Each, in fact, is a cell (p. 320) containing a nucleus; but to these essentials other structures—shell, white, yolk, etc.—may be added. Each egg, under proper conditions, is capable of growing into a form like that which produced it. The essential condition is that a peculiarly modified cell, the **spermatozoan**, unites with the egg, and then the compound cell is capable of development.† Reduced to its simplest terms, the process of development may be briefly stated thus:

After union with the sperm cell **(fertilization)** the egg divides again and again, the result being the formation of a large number of cells, all connected together, which later arrange themselves in layers (p. 318), and then develop into organs. This type of reproduction is known as sexual reproduction, since egg-cell and sperm-cell are produced by animals of different sexes.

---

* The gills, kidneys, and heart of the Chitons (p. 273) and the Nautilus (p. 267) are supposed to present indications of segmentation.

† In a few cases, as in the honey-bee, the eggs can develop without union with a spermatozoan.

In many Protozoa something similar occurs. Here we find a union of different individuals, and as each protozoan is a single cell, this union of individuals is comparable, to a certain extent, to the union of egg-cell and sperm-cell. With the Protozoa, however, after this union (conjugation) the individuals separate and each divides, thus producing new individuals (cells), which differ from the cells produced by the division of the egg in that they never arrange themselves into layers, but each forms a distinct individual like the parent.

Besides this sexual reproduction many animals possess the power of reproducing asexually. In these cases the animal may divide into two, or a small portion may protrude as a bud which will eventually produce an individual more or less like the parent. This asexual reproduction is very common among the cœlenterates, but it may also occur among the lower worms (p. 243), the Polyzoa, the tunicates, etc.

In many instances this asexual reproduction does not result in the formation of distinct and separate animals, but buds and parents may remain somewhat intimately connected with each other, the result being the formation of what are known as colonies, of which Pennaria may be taken as a type. Here we are met with a difficulty in the use of terms. We have spoken heretofore of *individuals ;* but is each zooid in a colony of Pennaria an individual, or is the colony itself to be so regarded, the zooids being regarded as organs ?

In many cases this reproduction by budding results in the formation of parts very different from each other. Thus in the hydroid (Fig. 147) abundant on shells inhabited by hermit-crabs, the colony consists of three different kinds of hydranths: (1) the feeding hydranths (*f*) which take

nourishment for the whole colony; (2) the protective hydranths (*p*) which lack mouths, but which are richly provided with nettle-cells; and (3) the reproductive hydranths
(*r*), the sole function of which is the reproduction of the
species. In the Siphonophores this differentiation is carried

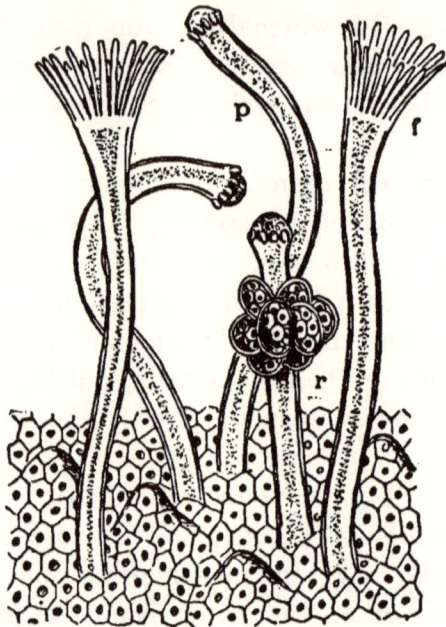

FIG. 147.—Part of a colony of the hydroid, *Hydractinia,* an illustration of
polymorphism. *f*, feeding individuals; *p*, protective individuals; *r*, reproductive individual.

still farther (p. 307), for here seven different forms may be
developed.   When there are but two different forms in the
history of the species it is called **dimorphic** (from the
Greek meaning two forms); if more than two, the species
is **polymorphic**.

Besides the di- or polymorphism produced by budding,
similar conditions may arise in other ways.   Thus frequently we find **sexual dimorphism**, in which the male and

female of the same species are greatly different in their appearance. An example of this is familiar in the can-
kerworm-moths, the male of which is winged, the female wingless. Again, we have to recognize a **seasonal dimorphism.** Thus certain butterflies produce several broods in a year. Those of the summer broods are so different from those which come from co-coons which have passed

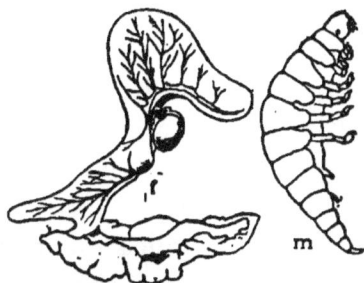

Fig. 148.—Male (*m*) and female (*f*) of one of the isopod Crustacea, an extreme example of sexual dimorphism.

through the winter, that without following through the whole history the relationships would not be suspected.

Closely connected with this polymorphism is the phenomenon of alternation of generations, of which instances are abundant in some groups of the animal kingdom (p. 307). Thus in the butterflies just mentioned, from the eggs of the winter-brood individuals are produced the summer brood presenting far different appearances from the parents, while the eggs of the summer brood produce in turn the winter brood. Again, in certain gall-wasps the difference between two generations is so great—both in appearance and in habits—that they would never be regarded as belonging to the same species, or even to the same genus, were it not that the whole history had been followed, so that it was ascertained that each generation resembles, not its parents, but its grandparents.

Many animals in the course of their development pass through a **metamorphosis,** which is not to be confused with polymorphism. In forms where a metamorphosis occurs the young, as it hatches from the egg, is greatly different from the parent, but by successive changes of form it at last reaches the adult condition in which it resembles closely the parent. These metamorphoses at times give us clues as to the past history of the group. Thus the larvæ of Echinoderms (p. 291) and the tadpoles of the Anura (p. 50) point to the fact that the first group has descended from markedly bilateral ancestors, and that the radiate condition of the adult has been secondarily acquired; while the history of the frog is evidence that these batrachians have sprung from tailed water-breathing ancestors. In the Insects, on the other hand, the larval and pupal stages have far less significance, but apparently have been introduced into the history the better to adapt these forms to the various conditions of their existence.

THE ANIMAL KINGDOM.

All of the different forms already enumerated—both protozoan and metazoan—are called animals,* in contrast with another great group of living forms, the plants, which form the Vegetable Kingdom.

At first sight the animals and the plants seem entirely distinct. We say that animals move, have sensation, have organs of feeding, of respiration, motion, etc., and that the plants lack all these. When we contrast a cat and a cabbage these and many other points of difference are at once forced upon us, while the features in which they resemble one another seem to be extremely few. When, however, we carry our comparisons farther and take the lower forms into account we soon find that these distinctions fail. We find many animals which are as firmly fixed as any tree, while we find many undoubted plants which move through the water as freely as any fish. We find, again, many plants which have undoubted powers of sensation. House-plants in a window turn their leaves towards the source of light; the leaves of the sensitive-plant droop if they be touched; while the reproductive elements (zoospores) of many low aquatic plants will recognize the presence of and swim towards a trace of malic acid. On the other hand, sensory organs are as poorly developed in sponges, and in many Protozoa, as in many plants.

* Frequently the term animal is restricted to members of the group of mammals. Thus we hear one say "animals and birds." This is not correct. A bird, a fish, or a clam is as truly an animal as a cat.

Plants really have their organs of feeding and of respiration in their roots and leaves, while animals as high as the parasitic worms have no organs for taking food or for respiration, the absorption of nourishment taking place at any spot on the surface of the body.

Several other tests have been suggested to separate animals from plants. Plants reproduce by seeds, by spores, and by buds; animals by means of eggs. Plants take up carbon dioxide and give off oxygen; animals use oxygen and give off carbon dioxide. Plants take either liquid or gaseous nourishment, while animals partake of solid food. Plants may have a peculiar green coloring substance called chlorophyl, lacking in animals. Plants produce a peculiar chemical substance known as cellulose. These features when accurately analyzed are all seen to have their exceptions. Many animals reproduce by budding, while the sexual reproduction of animals and of plants is essentially the same. Plants require oxygen as much as animals, and it is only the green plants which give off oxygen; a mushroom or a toadstool takes up oxygen and gives off carbon dioxide the same as does any animal. Quite a number of animals possess chlorophyl, while it is lacking from many plants; and cellulose is found even in the Tunicata. In the matter of food the distinction is a little sharper. While some animals like the parasitic worms take only nourishment in solution, no plant takes solid nourishment.

Yet, although we cannot frame a perfect definition which will at once separate all animals from all plants, we practically have little difficulty in deciding in any given case that is likely to arise in our every-day experience as to whether the form in question shall be placed in the one kingdom or in the other.

The difficulty of framing a definition arises from the fact

that both animals and plants are both members of the living world, and hence have many features in common, which may be summarized in the expression that both are alive. We do not know what life * is ; we only know it by the phenomena which it exhibits, which may be briefly stated as follows:

All living beings are composed of a peculiar substance (or group of substances) known as **protoplasm**, and this protoplasm is known only as the product of life. When unmixed with other substances it is semifluid, transparent, and slightly heavier than water. It contains a large number of chemical elements—carbon, oxygen, hydrogen, nitrogen, sulphur, and phosphorus predominating—but how these are arranged is as yet one of the mysteries. When treated with the reagents of the chemist it dies and is no longer protoplasm.

This protoplasm, and consequently the animals and plants which contain it, exhibits certain properties. It can take non-living substances and convert them into a part of itself, that is, make them alive. The bread and the roast beef which we eat are dead; yet we know that they become parts of ourselves, not in the shape of bread and roast beef, but as our own protoplasm. This process is known as assimilation, and continued assimilation results in growth. A snow-ball grows by accretions on the outside, but the growth of animals and plants occurs all through the body and throughout every part of it. It is a growth of the protoplasm.

Protoplasm has the power of spontaneous motion. Under

---

* Frequently the expression " vital force " is used, as if there were some distinct force in nature exhibiting itself only in living forms. This is entirely unnecessary, for each and every phenomenon of life can be explained by physical and chemical means,

favorable conditions we can see its particles changing their relative position, or we may see the mass move as a whole. It moves also in response to external influences, or, as the physiologist expresses it, it reacts to stimuli. Thus some protoplasm will turn to the light, other kinds will try to avoid it. Heat, up to a certain degree, will increase its action, while electricity will cause it to contract.

Protoplasm has the power of reproduction, by which we mean that portions can separate from the parent mass and can then carry on all the processes which could be performed before the separation took place.

These, and a number of other features not so easily described, are characteristic of protoplasm, and they occur in no non-living substance. These are, too, the phenomena of life, and hence protoplasm has been aptly termed the physical basis of life.

APPENDIX.

Alcohol.—The most important of all reagents. It can be purchased, tax-free, by incorporated institutions upon the fulfilment of certain conditions.* As it comes from the distiller it is usually about 95% alcohol, the rest being water. This is too strong for most purposes, and for the preservation of material it should be reduced to 70% by the addition of water.

Specimens for dissection should be kept in 70% alcohol, but in putting them up a weaker grade (50%) should be used first, and this should be changed every day or two at first. Plenty of alcohol should be used, otherwise the specimens will spoil.

Instead of alcohol for preservative purposes, other solutions are sometimes used with fairly good results. Among these may be mentioned:

Formol.—This is a 40% solution of formaldehyde, and for use this should be reduced by addition of water to a 2% solution (*i.e.*, 1 part formol to 49 of water), in which specimens may be kept in good condition for some months. The same care must be exercised as with alcohol to change the fluid frequently while hardening the specimens. Formol has the disadvantage of evaporating readily, and so the jars must be tightly sealed. It also has the disadvantage of freezing.

* These may be learned by application to the Collector of Internal Revenue in any district in the United States.

A second substitute for alcohol is **Wickersheimer's fluid.**
This is made by dissolving 100 grams of alum, 25 of common salt, 12 of saltpetre, 60 of potassic carbonate, and 20
of white arsenic (arsenious acid) in 3 litres of boiling
water.   To this, when cold, add 1200 grams of glycerine
and 300 of alcohol.   Change the specimens once or twice,
and keep them in at least twice their bulk of the fluid.

Injections are made as a means of more readily following
tubular structures, especially blood-vessels, and consist in
forcing into these tubes colored material which will render
them more easily recognized.   For many injections simple
apparatus may be used.   Thus frequently a glass tube
drawn out to a point can be filled with the injecting fluid
and then, when the end of the tube is inserted into the
blood-vessel, the fluid can be forced into the artery or vein
by the pressure of the breath.   It is, however, more satisfactory to use the regular injecting syringe, sold by all
dealers in naturalists' supplies.   These are provided with
small tubes (canulas) for insertion into the vessel to be injected, and these are grooved at the tip so that they may
be firmly tied into the artery or vein.

Most of the injections called for in the present work
can be made either through the aorta or through the ventricle.   The ventricle is cut open, and the canula is forced
through this opening into the aorta, around which a string
is passed and tied, thus holding the tube firmly in place.
The syringe is then filled with the injecting fluid (see
below) and connected with the canula, when a pressure
upon the piston will force the fluid into the blood-vessels.
Too much pressure should not be exerted, as the vessels are
liable to rupture.

Various injecting fluids have been proposed, but the
following are ample for all purposes, and they have, be-

sides, the advantage of not requiring heat, which in the case of some forms causes a softening of the walls of the blood-vessels.

Starch Injection Mass.—Grind together in a mortar one volume of dry starch, one of a 2½% aqueous solution of chloral hydrate, and one-fourth volume each of 95% alcohol and of the "color." The "color" consists of equal volumes of dry color (vermilion, chrome yellow, Prussian blue, etc.) glycerine and alcohol. The mixture will keep indefinitely, but requires thorough stirring before use and quick usage, as the starch and color settle rapidly.

Gum Injection Fluid.—Make a rather thick solution of gum arabic in water; color it with carmine or soluble Prussian blue, and strain through muslin. With the addition of a little thymol the fluid will keep well if tightly corked. After injection, place the animal in alcohol, which hardens the gum.

By using both injection masses in succession the complete circulatory system may be injected (double injection). To accomplish this, first inject with the gum fluid, colored blue, and then follow with the starch mass colored red. The gum will flow through the finest vessels, but the starch mass will stop at the capillaries.

Study of Vertebrate Brains.—If material be abundant, the study of the brain and its nerves will be much facilitated by putting heads of the various forms in the fluid mentioned below a week or two before the dissection is to take place. The fluid, which should be changed two or three times, softens (decalcifies) the bones, and at the same time hardens the nervous structures. It is composed of equal parts of 95% (commercial) alcohol and 10% nitric acid. The heads should be washed for an hour or two in water before dissection, as otherwise the acid will attack the dissecting instruments.

Fuchsin is one of the most easily used stains. It is made by dissolving one part of the aniline dye in two hundred parts of water.

Picrosulphuric Acid is used for killing many animals without distortion. It is made by dissolving picric acid in water until no more will be taken up, and then adding to one hundred parts of the solution two parts of sulphuric acid. It is allowed to stand a day, is filtered, and is prepared for use by adding three parts of water to one of the stock solution. Specimens killed in this fluid are stained yellow, and should be washed in several changes of water before being placed in alcohol or formol. It takes from one to three hours to kill.

INDEX.

Jaws, 131
Jelly-fishes, 304, 305, 307
Jugular veins, 61, 90
Jumping mice, 108
June-bug, 186
June-bug, dissection of, 177

Kangaroo, 104
Katydid, 183
Keel of sternum, 74
Kidney, 12, 90
Kingfishers, 83

Labium, 173
Labrum, 173
Lac, 198
Lacertilia, 67
Ladybugs, 185
Lampreys, 151
Lamp-shells, 247
Lancelets, 153
Larva, 49, 184
Larynx, 92
Lateral line, 10, 137
Leaf-beetles, 186
Leaf-hoppers, 197
Leeches, 240
Lemming, 108
Lemurs, 123
Lens, 139
Leopard, 121
Lepas, 211
Lepidoptera, 197, 222
Lepidosteus, 39
Leptocardii, 153
Leucania, 199
Leucocytes, 149
Lice, 198
Lice, fish, 210
Lice, plant, 197
Life, 339
Limpets, 275
Lines of growth, 248, 270
Lion, 121
Lingual ribbon, 255, 271
Liver, 142
Liver-fluke, 244
Liver-rot, 244
Lizards, 67

Llama, 118
Lobster, 164
Lobster, dissection of, 157
Locusts, 183
Loon, 78
Lophobranchii, 32
Lumbar region, 130
Lung-fishes, 39
Lungs, 50
Lymphatics, 88
Lymph-heart, 40
Lynx, 121

Macaques, 124
Mackerel, 29
Macrura, 164
Madreporite, 278
Maggots, 222
Malacostraca, 209
Malpighian tubes, 174, 215
Mammalia, 97
Man, 124
Manatee, 113
Mandibulatæ, 216
Mandible, 159
Mandibles of birds, 59
Mangabey, 124
Manis, 106
Mantle, 248, 269
Marabou, 80
Marmosets, 123
Marsupialia, 103
Marten, 121
Mastodon, 115
Maxillæ, 158
Maxillary, 10
Maxillipeds, 158
May-flies, 219
Measuring-worm, 202
Mediastinum, 90
Medusa-buds, 300
Medusæ, 304
Medulla oblongata, 15, 135
Megatherium, 106
Menhaden, 26
Merostomata, 229
Mesenterial artery, 88
Mesenterial filaments, 298
Mesenterial vein, 88
Mesenteries of sea-anemone, 297

Hertwig's General Principles of Zoology. From the Third Edition of Dr. Richard Hertwig's *Lehrbuch der Zoologie.* Translated and edited by GEORGE WILTON FIELD, Professor in Brown University. 226 pp. 8vo. $1.60 *net.*

Howell's Dissection of the Dog. As a Basis for the Study of Physiology. By W. H. HOWELL, Professor in the Johns Hopkins University. 100 pp. 8vo. $1.00 *net.*

Jackman's Nature Study for the Common Schools. (Arranged by the Months.) By WILBUR JACKMAN, Teacher of Natural Science, Cook County Normal School, Chicago, Ill. 448 pp. 12mo. $1.20 *net.*

Kerner & Oliver's Natural History of Plants. Translated by Prof. F. W. OLIVER, of University College, London. 4to. 4 parts. With over 1000 illustrations and 16 colored plates. $15.00 *net.*

Macalister's Zoology of the Invertebrate and Vertebrate Animals. By ALEX. MACALISTER. Revised by A. S. PACKARD. 277 pp. 16mo. 80 cents *net.*

MacDougal's Experimental Plant Physiology. On the Basis of Oels' *Pflanzenphysiologische Versuche.* By D. T. MACDOUGAL, University of Minnesota. vi + 88 pp. 8vo. $1.00 *net.*

Macloskie's Elementary Botany. With Students' Guide to the Examination and Description of Plants. By GEORGE MACLOSKIE, D.Sc., LL.D. 373 pp. 12mo. $1.30 *net.*

McMurrich's Text-book of Invertebrate Morphology. By J. PLAYFAIR MCMURRICH, M.A., Ph.D., Professor in the University of Cincinnati. vii + 661 pp. 8vo. *New Edition.* $3.00 *net.*

McNab's Botany. Outlines of Morphology, Physiology, and Classification of Plants. By WILLIAM RAMSAY MCNAB. Revised by Prof. C. E. BESSEY. 400 pp. 16mo. 80c. *net.*

Martin's The Human Body. See American Science Series.

***Merriam's Mammals of the Adirondack Region,** Northeastern New York. With an Introductory Chapter treating of the Location and Boundaries of the Region, its Geological History, Topography, Climate, General Features, Botany, and Faunal Position. By Dr. C. HART MERRIAM. 316 pp. 8vo. $3.50 *net.*

Newcomb & Holden's Astronomies. See American Science Series.

***Noel's Buz;** or, The Life and Adventures of a Honey Bee. By MAURICE NOEL. 134 pp. 12mo. $1.00.

Noyes's Elements of Qualitative Analysis. By WM. A. NOYES, Professor in the Rose Polytechnic Institute. 91 pp. 8vo. 80c. *net.*

Packard's Entomology for Beginners. For the use of Young Folks, Fruit-growers, Farmers, and Gardeners. By A. S. PACKARD. xvi + 367 pp. 12mo. *Third Edition, Revised.* $1.40 *net.*

Postage 8 per cent additional on net books. Descriptive list free.

Packard's Guide to the Study of Insects, and a Treatise on those Injurious and Beneficial to Crops. For Colleges, Farm-schools, and Agriculturists. By A. S. Packard. With 15 plates and 670 wood-cuts. *Ninth Edition.* 715 pp. 8vo. $4.50 *net.*

—— **Outlines of Comparative Embryology.** By A. S. PACKARD. Copiously illustrated. 243 pp. 8vo. $2.00 *net.*

—— **Zoologies.** See American Science Series.

Perkins's Outlines of Electricity and Magnetism. By CHAS. A. PERKINS, Professor in the University of Tennessee. 277 pp. 12mo. $1.10 *net.*

Pierce's Problems in Elementary Physics. Chiefly numerical. By E. DANA PIERCE, of the Hotchkiss School. 194 pp. 12mo. 60c. *net.*

Price's The Fern Collector's Handbook and Herbarium. By Miss SADIE F. PRICE. With 72 plates. (*November*, 1896.)

Remsen's Chemistries. See American Science Series.

Scudder's Butterflies. By SAMUEL H. SCUDDER. x + 322 pp. 12mo. $1.20 *net.*

—— **Brief Guide to the Commoner Butterflies.** By SAMUEL H. SCUDDER. xi + 206 pp. 12mo. $1.00 *net.*

—— **The Life of a Butterfly.** A Chapter in Natural History for the General Reader. By S. H. SCUDDER. 186 pp. 16mo. 80c. *net.*

Sedgwick and Wilson's Biology. See American Science Series.

*Step's Plant Life.** By EDWARD STEP. Popular Papers on the Phenomena of Botany. 12mo. 148 Illustrations. $1.00 *net.*

Underwood's Our Native Ferns and their Allies. By LUCIEN M. UNDERWOOD, Professor in DePauw University. *Revised.* 156 pp. 12mo. $1.00 *net.*

Williams's Elements of Crystallography. By GEORGE HUNTINGTON WILLIAMS, late Professor in the Johns Hopkins University. x + 270 pp. 12mo. *Revised and Enlarged.* $1.25 *net.*

Williams's Geological Biology. An Introduction to the Geological History of Organisms. By HENRY S. WILLIAMS, Professor of Geology in Yale College. 8vo. 395 pp. $2.80 *net.*

Woodhull's First Course in Science. By JOHN F. WOODHULL, Professor in the Teachers' College, New York City.
I. Book of Experiments. xiv + 79 pp. 8vo. Paper. 50c. *net.*
II. Text-Book. xv + 133 pp. 12mo. Cloth. 65c. *net.*
III. Box of Apparatus. $2.00 *net* (*actual cost to the publishers*).

Zimmermann's Botanical Microtechnique. Translated by JAMES ELLIS HUMPHREY, S.C. xii + 296 pp. 8vo. $2.50 *net.*

HENRY HOLT & CO., 29 WEST 23D ST., NEW YORK.

July, 1897.

SCIENCE
REFERENCE AND TEXT-BOOKS

PUBLISHED BY

HENRY HOLT & COMPANY,

29 WEST 23D STREET, NEW YORK.

Books marked * are chiefly for reference and supplementary use, and are to be found in Henry Holt & Co.'s List of Works in General Literature. For further particulars about books not so marked see Henry Holt & Co.'s Descriptive Science Catalogue. Either list free on application. Excepting JAMES' PSYCHOLOGIES *and* WALKER'S POLITICAL ECONOMIES, *both in the American Science Series, this list contains no works in Philosophy or Political Economy. Postage on net books 8 per cent. additional.*

American Science Series

1. Astronomy. By SIMON NEWCOMB, Professor in the Johns Hopkins University, and EDWARD S. HOLDEN, Director of the Lick Observatory, California.
Advanced Course. 512 pp. 8vo. $2.00 *net.*
The same. *Briefer Course.* 352 pp. 12mo. $1.12 *net.*

2. Zoology. By A. S. PACKARD, JR., Professor in Brown University. *Advanced Course.* 722 pp. 8vo. $2.40 *net.*
The same. *Briefer Course.* 338 pp. 12mo. $1.12 *net.*
The same. *Elementary Course.* 290 pp. 12mo. 80 cents *net.*

3. Botany. By C. E. BESSEY, Professor in the University of Nebraska. *Advanced Course.* 611 pp. 8vo. $2.20 *net.*
The same. *Briefer Course.* (*Entirely new edition*, 1896.) 356 pp. 12mo. $1.12 *net.*

4. The Human Body. By H. NEWELL MARTIN, sometime Professor in the Johns Hopkins University.
Advanced Course. (*Entirely new edition*, 1896.) 685 pp. 8vo. $2.50 *net.*
Copies without chapter on Reproduction sent when specially ordered.
The same. *Briefer Course.* 377 pp. 12mo. $1.20 *net.*
The same. *Elementary Course.* 261 pp. 12mo. 75 cents *net.*
The Human Body and the Effect of Narcotics. 261 pp. 12mo. $1.20 *net.*

5. Chemistry. By IRA REMSEN, Professor in Johns Hopkins University.
Advanced Course (Inorganic). 850 pp. 8vo. $2.80 *net.*
The same. *Briefer Course.* (*Entirely new edition*, 1893.) 435 pp. $1.12 *net.*
The same. *Elementary Course.* 272 pp. 12mo. 80 cents *net.*
Laboratory Manual (*to Elementary Course*). 196 pp. 12mo. 40 cents, *net.*
Chemical Experiments. By Prof. REMSEN and Dr. W. W. RANDALL. (*For Briefer Course.*) *No* blank pages for notes. 158 pp 12mo. 50c. *net.*

6. Political Economy. By FRANCIS A. WALKER, President Massachusetts Institute of Technology. *Advanced Course.* 537 pp. 8vo. $2.00 *net.*
The same. *Briefer Course.* 415 pp. 12mo. $1.20 *net.*
The same. *Elementary Course.* 423 pp. 12mo. $1.00 *net.*

7. General Biology. By Prof. W. T. SEDGWICK, of Massachusetts Institute of Technology, and Prof. E. B. WILSON, of Columbia College. (*Revised and enlarged*, 1896.) 231 pp. 8vo. $1.75 *net.*

8. Psychology. By WILLIAM JAMES, Professor in Harvard College. *Advanced Course.* 689 + 704 pp. 8vo. 2 vols. $4.80 *net.*
The same. *Briefer Course.* 478 pp. 12mo. $1.60 *net.*

9. Physics. By GEORGE F. BARKER, Professor in the University of Pennsylvania. *Advanced Course.* 902 pp. 8vo. $3.50 *net.*

10. Geology. By THOMAS C. CHAMBERLIN and ROLLIN D. SALISBURY, Professors in the University of Chicago. (*In Preparation.*)

Allen's Laboratory Exercises in Elementary Physics. By CHARLES R. ALLEN, Instructor in the New Bedford, Mass., High School. *Pupils' Edition:* x + 209 pp. 12mo. 80c. *net. Teachers' Edition:* $1.00 *net.*

Arthur, Barnes, and Coulter's Handbook of Plant Dissection. By J. C. ARTHUR, Professor in Purdue University, C. R. BARNES, Professor in University of Wisconsin, and JOHN M. COULTER, President of Lake Forest University. xi + 256 pp. 12mo. $1.20 *net.*

Barker's Physics. See American Science Series.

Beal's Grasses of North America. For Farmers and Students. By W. J. BEAL, Professor in the Michigan Agricultural College. 8vo. Copiously illustrated. 2 vols. Vol. I., 457 pp. $2.50 *net.* Vol. II., 707 pp. $5.00 *net.*

Bessey's Botanies. See American Science Series.

Black and Carter's Natural History Lessons. By GEORGE ASHTON BLACK, Ph.D., and KATHLEEN CARTER. (For very young pupils.) x + 98 pp. 12mo. 50c. *net.*

Bumpus's Laboratory Course in Invertebrate Zoology. By HERMON C. BUMPUS, Professor in Brown University, Instructor at the Marine Biological Laboratory, Wood's Holl, Mass. *Revised.* vi + 157 pp. 12mo. $1.00 *net.*

Cairns's Quantitative Chemical Analysis. By FREDERICK A. CAIRNS. Revised and edited by Dr. E. WALLER. 417 pp. 8vo. $2.00 *net.*

Champlin's Young Folks' Astronomy. By JOHN D. CHAMPLIN, Jr., Editor of *Champlin's Young Folks' Cyclopædias.* Illustrated. vi + 236 pp. 16mo. 48c. *net.*

Crozier's Dictionary of Botanical Terms. By A. A. CROZIER. 202 pp. 8vo. $2.40 *net.*

Hackel's The True Grasses. Translated from "Die natürlichen Pflanzenfamilien" by F. LAMSON-SCRIBNER and EFFIE A. SOUTHWORTH. v + 228 pp. 8vo. $1.50.

Hall's First Lessons in Experimental Physics. For young beginners, with quantitative work for pupils and lecture-table experiments for teachers. By EDWIN H. HALL, Assistant Professor in Harvard College. viii + 120 pp. 12mo. 65c. *net.*

Hall and Bergen's Text-book of Physics. By EDWIN H. HALL, Assistant Professor of Physics in Harvard College, and JOSEPH Y. BERGEN, Jr., Junior Master in the English High School, Boston. xviii + 388 pp. 12mo. $1.25 *net.*

PSYCHOLOGY
ETHICS AND PHILOSOPHY
REFERENCE AND TEXT-BOOKS

PUBLISHED BY

HENRY HOLT & COMPANY
29 WEST 23D STREET, NEW YORK

Books marked * are chiefly for reference and supplementary use, and to be found in Henry Holt & Co.'s List of Works in General Literature. For further particulars about books not so marked see Henry Holt & Co.'s Descriptive Educational Catalogue. Either list free on application. Postage on net books 10 per cent. additional.

*Bain's John Stuart Mill. A Criticism with Personal Recollections. By Prof. ALEXANDER BAIN of Aberdeen. 12mo. 214 pp. $1.00.

*——James Mill. A Biography. With portrait. 12mo. 498 pp. $2.00.

Baldwin's Handbook of Psychology. By Prof. JAMES MARK BALDWIN of Toronto. 2 vols. (sold separately). 8vo.
Vol. I. Senses and Intellect. 357 pp. $1.80, *net.*
Vol. II. Feeling and Will. 406 pp. $2.00, *net.*

—— Elements of Psychology. 372 pp. 12mo. $1.50, *net.*

Descartes : The Philosophy of Descartes. Selected and translated by Prof. H. A. P. TORREY of the University of Vermont. (*Sneath's Modern Philosophers.*) xii + 345 pp. 12mo. $1.50, *net.*

Falckenberg's Modern Philosophy from Nicolas of Cusa to the Present Time. By Prof. RICHARD FALCKENBERG of Erlangen. Translated with the author's co-operation by Prof. A. C. ARMSTRONG, Jr., of Wesleyan. 665 pp. 8vo. $3.50, *net.*

Hegel : Philosophy of Hegel. Translated extracts from Hegel's works, with an introduction by Prof. JOSIAH ROYCE of Harvard. (*Sneath's Modern Philosophers.*) (*In preparation.*)

*Hillebrand's German Thought. From the Seven Years' War to Goethe's Death. Six Lectures delivered at the Royal Institution of Great Britain. By CARL HILLEBRAND. 12mo. 306 pp. $1.75.

*Holland's Rise of Intellectual Liberty, from Thales to Copernicus. A History. By FREDERICK MAY HOLLAND. 8vo. 458 pp. $3.50.

Hume : The Philosophy of Hume. Extracts from the "Treatise of Human Nature." With an introduction by HERBERT A. AIKINS, Professor in Western Reserve University. 176 pp. 12mo. (*Sneath's Modern Philosophers.*) $1.00, *net.*

Hyde's Practical Ethics. By Pres. WM. DE WITT HYDE of Bowdoin. 12mo. 219 pp. 80 cents, *net.*

James' Principles of Psychology. (*American Science Series.*) *Advanced Course.* By Prof. WM. JAMES of Harvard. 2 vols. 8vo. 701 + 710 pp. $4.80, *net.*

—— **Psychology.** *Briefer Course.* 12mo. 491 pp. $1.60, *net.*

Jastrow's Chapters in Modern Psychology. By Prof. JOSEPH JASTROW of the University of Wisconsin. (*In preparation.*)

Kant : The Philosophy of Kant. Selected and translated by Prof. JOHN WATSON of Queen's College, Canada. x + 356 pp. 12mo. (*Sneath's Modern Philosophers.*) $1.75, *net.*

Locke : The Philosophy of Locke. Extracts from the "Essay concerning Human Understanding." Edited by Prof. JOHN E. RUSSELL of Williams College. 160 pp. 12mo. (*Sneath's Modern Philosophers.*) $1.00, *net.*

*****Martineau's Essays, Philosophical and Theological.** By JAMES MARTINEAU. 2 vols. 8vo. 428 + 438 pp. $5.00.

*****Maude's The Foundation of Ethics.** By JOHN EDWARD MAUDE. Edited by Prof. WM. JAMES of Harvard. 12mo. 224 pp. $1.50.

*****Mill's Three Essays on Religion, and Berkeley.** By JOHN STUART MILL. 8vo. 313 pp. $2.00.

***—— The Autobiography.** 8vo. 319 pp. $2.00.

***—— Dissertations and Discussions.** 5 vols. 8vo. $2.00 per vol.

***—— Examination of Sir William Hamilton's Philosophy.** 8vo. 354 pp. $2.75.

***—— Comte's Positive Philosophy.** 8vo. 182 pp. $1.25.

*****Mill, John Stuart : His Life and Works.** Twelve sketches by HERBERT SPENCER, HENRY FAWCETT, FREDERIC HARRISON, and other distinguished authors. 16mo. 96 pp. $1.00.

*****Nicholls' The Psychology of Time.** By HERBERT NICHOLLS, Fellow of Clark. 8vo. 140 pp. $1.50, *net.*

Paulsen's Introduction to Philosophy. By Prof. FRIEDRICH PAULSEN of Berlin. Translated with the author's sanction by Prof. FRANK THILLY of the University of Missouri. With a preface by Prof. WM. JAMES of Harvard. xxiv + 437 pp. 8vo. $3.50, *net.*

Reid : The Philosophy of Reid. The "Inquiry into the Human Mind on the Principles of Common Sense." Edited by Dr. E. HERSHEY SNEATH of Yale. viii + 368 pp. 12mo. (*Sneath's Modern Philosophers.*) $1.50, *net.*

Spinoza : The Philosophy of Spinoza. Parts I, II, V of the "Ethics," and Extracts from III, IV. Translated and edited by Prof. GEO. STUART FULLERTON of the University of Pennsylvania. *Second Edition, Enlarged.* vi + 358 pp. 12mo. (*Sneath's Modern Philosophers.*) $1.50, *net.*

Zeller's Outlines of the History of Greek Philosophy. By Dr. EDWARD ZELLER. Translated with the author's sanction by SARAH F. ALLEYNE and EVELYN ABBOTT. 12mo. 377 pp. $1.40, *net.*

Postage on net books 8 per cent additional. Descriptive list free.

HENRY HOLT & CO., 29 WEST 23D ST., NEW YORK.
September, 1895.

SOME ENGLISH TRANSLATIONS

About, Edmond : The Man with the Broken Ear. 16mo. $1.—The Notary's Nose. 16mo. $1.

Auerbach, B.: The Villa on the Rhine. (Davis.) 2 vols. 16mo. $2. —On the Heights. (Stern.) 2 vols. 16mo. $2.—On the Heights. (Bunnett). 16mo. Paper. 1 vol. 30c.

Bacourt, Chevalier de : Souvenirs of a Diplomat. 12mo. $1.50.

Berlioz, Hector : Selections from his Letters and Writings. (Apthorp.) 12mo. $2.

Brink, Bernhard ten: English Literature. Vol. I. (Kennedy.) Large 12mo. $2. Vol. II. (Robinson.) Large 12mo. $2.— Five Lectures on Shakespeare. (Franklin.) 12mo. $1.25.

Chevrillon, André: In India. (Marchant) 12mo. $1.50.

Falke, Jakob von: Greece and Rome: Their Life and Art. (Browne.) Quarto. $10.

Firdusi : The Epic of Kings. (Zimmern.) 12mo. $2.50.

Gautier, Léon: Chanson de Roland. (Rabillon). 12mo. $1.25.

Gautier, Théophile : A Winter in Russia. (Ripley.) 12mo. $1.75.— Constantinople. (Gould.) 12mo. $1.75.

Gavard, Charles : A Diplomat in London. (Hodder.) 12mo.

Goethe, J. W. von : Poems and Ballads. (Gibson.) (*Library of Foreign Poetry.*) 16mo. $1.50.

Guérin, Maurice de : Journal. (Fisher.) 12mo. $1.25.

Guyau, Jean Marie : The Irreligion of the Future. (Hodder.)

Heine, Heinrich : Book of Songs. (Leland.) (*Library of Foreign Poetry.*) 16mo. 75c.—The Romantic School. (Fleishman). 12mo. $1.50.—Life Told in His Own Words. (Dexter.) 12mo. $1.75.

Hertz, Henrik : King René's Daughter. (Martin.) (*Library of Foreign Poetry.*) 16mo. $1.25.

Heyse, Paul : The Children of the World. 12mo. $1.25.

Kalevala. Selections. (Porter.) 16mo. $1.50.

Kalidasa : Shakuntala. (Edgren.) (*Library of Foreign Poetry.*) 16mo. $1.50.

Klaczko, Julian : Rome and the Renaissance. (Marchant.)

Knortz, Karl: Representative German Poems. 12mo. $2.50.

Lessing, G. E.: Nathan the Wise. (Frothingham.) (*Library of Foreign Poetry.*) 16mo. $1.50.

Lockhart, J. G.: Ancient Spanish Ballads. (*Library of Foreign Poetry.*) 16mo. $1.25.

Moscheles, Ignatz : Recent Music and Musicians. (A. D. Coleridge.) 12mo. $2.

Roumanian Fairy Tales. (Percival.) 12mo. $1.25.

Rousselet, Louis : Ralph, the Drummer Boy. (Gordon.) 12mo. $1.50.

Rydberg, Victor : Magic of the Middle Ages. (Edgren.) 12mo. $1.50.

Sainte-Beuve, C. A.: English Portraits. 12mo. $2.

Spielhagen, Frederick : Problematic Characters. (de Vere.) 16mo. Paper. 50 cents.—Through Night to Light. (de Vere.) 16mo. Paper. 50 cents.—The Hohensteins. (de Vere.) 16mo. Paper. 50 cents.—Hammer and Anvil. (Browne.) 16mo. Paper. 50 cents.

Sylva, Carmen : Pilgrim Sorrow. 16mo. $1.25.

Taine, H. A.: Italy, Rome and Naples. (Durand.) Large 12mo. $2.50.—Italy, Florence and Venice. (Durand.) Large 12mo. $2.50.—Notes on England. (Rae.) Large 12mo. $2.50.—A Tour through the Pyrenees. (Fiske.) Large 12mo. $2.50.—Notes on Paris. (Stevens.) Large 12mo. $2.50.—History of English Literature. (Van Laun.) 2 vols. Large 12mo. $5. The same. 12mo. 1 vol. $1.25. The same. Large 12mo. 4 vols. *In press.* —On Intelligence. (Haye.) 2 vols. Large 12mo. $5.—Lectures on Art. *First Series.* (Durand.) Large 12mo. $2.50.—Lectures on Art. *Second Series.* (Durand.) Large 12mo. $2.50.—The Ancient Régime. (Durand.) Large 12mo. $2.50.—The French Revolution. (Durand.) 3 vols. $7.50.—The Modern Régime. Vol. I. Large 12mo. $2.50.—The Modern Régime. Vol. II. Large 12mo. $2.50.

Tegnér, Esaias : Frithiof's Saga. (Blackley.) 16mo. $1.50.

Wagner, Wilhelm Richard : Art Life and Theories of Richard Wagner. (Burlingame.) 12mo. $2.—Ring of the Nibelung. (Dippold.) 12mo. $1.50.

Witt, C.: Classic Mythology. (Younghusband.) 12mo. * $1.

* denotes net price.

HENRY HOLT & CO., NEW YORK.